GROUND STUDIES FOR PILOTS

RADIO AIDS

Related titles on the JAR syllabus

Ground Studies for Pilots series

Radio Aids
R.B. Underdown & David Cockburn
0-632-05573-1

Navigation
R.B. Underdown & Tony Palmer
0-632-05333-X

Flight Planning
R.B. Underdown & Anthony Stevens
0-632-05939-7

Meteorology
R.B. Underdown & Anthony Stevens
0-632-03751-2

Aviation Law for Pilots
Tenth Edition
R.B. Underdown & Tony Palmer
0-632-05335-6

Human Performance and Limitations in Aviation
Second Edition
R.D. Campbell & M. Bagshaw
0-632-04986-3

Aircraft Performance Theory for Pilots
P.J. Swatton
0-632-05569-3

GROUND STUDIES FOR PILOTS

RADIO AIDS

Sixth Edition

R.B. Underdown and **David Cockburn**

Blackwell Science

Sixth edition © Estate of R.B. Underdown and
Blackwell Science Ltd 2001

Blackwell Science Ltd
Editorial Offices:
Osney Mead, Oxford OX2 0EL
25 John Street, London WC1N 2BS
23 Ainslie Place, Edinburgh EH3 6AJ
350 Main Street, Malden
   MA 02148 5018, USA
54 University Street, Carlton
   Victoria 3053, Australia
10, rue Casimir Delavigne
   75006 Paris, France

Other Editorial Offices:

Blackwell Wissenschafts-Verlag GmbH
Kurfürstendamm 57
10707 Berlin, Germany

Blackwell Science KK
MG Kodenmacho Building
7–10 Kodenmacho Nihombashi
Chuo-ku, Tokyo 104, Japan

Iowa State University Press
A Blackwell Science Company
2121 S. State Avenue
Ames, Iowa 50014-8300, USA

First published in Great Britain by Crosby
Lockwood & Son Ltd 1970
Second edition 1974
Third edition 1979
Fourth edition published by Collins Professional
and Technical Books 1986
Fifth edition published by Blackwell Science 1993
Sixth edition 2001

Set in 10/13pt Palatino
by DP Photosetting, Aylesbury, Bucks
Printed and bound in Great Britain by
MPG Books Ltd, Bodmin, Cornwall

The Blackwell Science logo is a trade mark of
Blackwell Science Ltd, registered at the United
Kingdom Trade Mark Registry

DISTRIBUTORS

Marston Book Services Ltd
PO Box 269
Abingdon
Oxon OX14 4YN
(*Orders:*   Tel: 01235 465500
            Fax: 01235 465555)

USA and Canada
   Iowa State University Press
   A Blackwell Science Company
   2121 S. State Avenue
   Ames, Iowa 50014-8300
   (*Orders:*   Tel: 800-862-6657
               Fax: 515-292-3348
               Web: www.isupress.com
               email: orders@isupress.com

Australia
   Blackwell Science Pty Ltd
   54 University Street
   Carlton, Victoria 3053
   (*Orders:*   Tel: 03 9347 0300
               Fax: 03 9347 5001)

A catalogue record for this title is available from the
British Library

ISBN 0-632-05573-1

Library of Congress
Cataloging-in-Publication Data
Underdown, R.B.
   Ground studies for pilots. Radio aids/R.B.
Underdown. — 6th ed.
      p. cm.
   Added t.p. title: Ground studies for pilots
series. Radio aids.
   Includes index.
   ISBN 0-632-05573-1
   1. Radio in aeronautics.   I. Title: Ground
studies for pilots series. Radio aids.   II. Title.

TL693 .U49 2001
629.132′51 — dc21

                                        00-068878

For further information on
Blackwell Science, visit our website:
www.blackwell-science.com

Contents

Preface

The new edition of the classic textbook, originally written by Taylor and Parmar and revised by Roy Underdown, has been made necessary by the changeover to the European Joint Aviation Authorities' syllabuses for commercial and airline transport pilots' licences. The series *Ground Studies for Pilots* has been a major source of information for pilots for many years. However, as time has passed, the subject has developed. New equipment has been introduced into service, and new technology has revolutionised some older systems. Some older systems have assumed less importance, and some have even faded away into disuse. It is intended to continue to provide essential information not only for those studying for the JAA (and other) professional pilots' examinations, but also to give general guidance to those pilots attempting to keep up-to-date with modern developments in the field of radio aids to navigation and safety.

I have taken the opportunity presented by this revision to re-write much of the material in an attempt to reflect the changing importance of the various topics in a readable style. This volume covers the whole range of radio aids from the properties and propagation of radio waves to the equipment available to pilots in a modern cockpit. I have continued to include terrain and collision warning systems in this volume, even though they appear in different sections of the JAA examinations, because they use the principles of radio waves, and progress naturally from other aids already covered.

In a departure from tradition, most of the chapters end with summaries. These summaries are meant to act as aides-mémoire to understanding, not to substitute for the main text. Most chapters also include sample questions with which the reader can check understanding, and the answers are given at the back of the book. Specimen questions for the JAA examinations are available from the Civil Aviation Authority.

In addition to those companies and individuals whose help made previous editions possible, the following have given much appreciated assistance in providing technical assistance and illustrations for this edition.

- Rockwell-Collins (UK) Avionics
- Honeywell (formerly AlliedSignal Avionics)

- Meggitt Avionics
- BFGoodrich Aerospace

I must thank Brian Kendal, author of *Manual of Avionics*, whose book has provided several of my illustrations; Figure 4.2 was provided by Fernau Electronics, Figure 9.3 by Cossor. I should also like to express my thanks to those of my colleagues in the UK and abroad who have given me their time and provided advice.

David Cockburn
Gatwick

List of Abbreviations

This list includes some abbreviations which are not used in this volume, but which readers may meet in the field of radio navigation.

a/c	aircraft
a.c.	alternating current
ACARS	aircraft communications addressing and reporting system
ACAS	aircraft collision avoidance system
ACC	Area Control Centre (air traffic control)
AD	aerodrome (section of the AIP)
ADC	air data computer
ADF	automatic direction finder
ADI	attitude direction indicator
ADS	automatic data surveillance
ADS-B	automatic data surveillance – broadcast
AF	audio frequency
AFDS	autopilot flight director system
AFIS	aerodrome flight information service (air traffic service)
agl	above ground level
AHARS	attitude, heading and reference system
AI	attitude indicator
AIC	aeronautical information circular
AIP	Aeronautical Information Publication
AIRAC	amendments to the AIP
ALT	altimeter (mode C transponding in SSR) *or* alternate
AM	amplitude modulation
amsl	above mean sea level
ANO	Air Navigation Order (UK law)
ANT	antenna (sense antenna in ADF)
AoA	angle of attack
apch	approach
ARINC	Aeronautical Radio Inc. (an organisation of operators which agrees equipment standards)
ASDE	airport surface detection equipment (radar)
ATC	air traffic control

ATCC	Air Traffic Control Centre
ATS	air traffic service
AWR	airborne weather radar
Bcn	beacon
BFO	beat frequency oscillator
BIT	binary digit
BITE	built-in test equipment
brg	bearing
B-RNAV	basic area navigation system
C	used to denote the speed of electromagnetic waves in a vacuum
CA	coarse acquisition (GPS)
CAA	Civil Aviation Authority
Cat	category
CDI	course deviation indicator
CDU	control and display unit
CFIT	controlled flight into terrain
chan	channel
cm	centimetre(s)
COSPAS	search and rescue satellite system (Russian)
CPU	central processing unit
CRS	course
CRT	cathode-ray tube
c/s	callsign
CW	continuous wave *or* carrier wave
DA	drift angle *or* decision altitude
dB	decibels (unit of signal strength)
d.c.	direct current
DDM	difference in depth of modulation (ILS)
dev	deviation (magnetic)
DF	direction finding
D/F	direction finding
DFT	drift
DGPS	differential Global Positioning System
DH	decision height
DI	direction indicator
diff	difference/differential
dir	direction
dist	distance
DME	distance measuring equipment
DME/P	precision DME (MLS)

DOC	designated operational coverage (VOR)
DoD	Department of Defense (US)
DR	deduced reckoning (also known as dead reckoning)
DVOR	Doppler VOR

°E	degrees east of the prime meridian
EADI	electronic attitude direction indicator
EFIS	electronic flight instrument system
EGNOS	European geostationary navigation overlay service (satellite navigation)
EGPWS	enhanced ground proximity warning system
EHF	extremely high frequency (30–300 GHz)
EHSI	expanded horizontal situation indicator *or* electronic horizontal situation indicator
ELF	extremely low frequency
ENR	en-route (section of the AIP)
EPIRB	emergency position indicating rescue beacon
ETA	estimated time of arrival
EXT	external

FAA	Federal Aviation Administration (US)
FAF	final approach fix
FAWP	final approach waypoint
FBO	fixed base operator (maintenance organisation)
f_d	Doppler shift in frequency
FDE	failure detection and exclusion (GPS)
FDSU	flight data storage unit (FMS)
FL	flight level
FM	frequency modulation *or* fan marker
FMC	flight management computer
FMCW	frequency modulated continuous wave
FMS	flight management system
FPL	flight plan
ft	feet
ft/s	feet per second

g	acceleration due to the earth's gravity (or its equivalent value)
GCA	ground controlled approach
GDOP	geometric dilution of position
GHz	gigahertz (10^9 Hertz)
GLONASS	global orbiting navigation satellite system (Russian)
GNM	ground nautical miles
GNSS	global navigation satellite system (generic term)

GP	glidepath
GPS	global positioning system (US)
GPWS	ground proximity warning system
GRI	group repetition interval (LORAN)
GS	ground speed *or* glideslope; G/S on instrumentation
Hdg	heading
Hdg (M)	magnetic heading
Hdg (T)	true heading
HF	high frequency (3–30 MHz)
HFDL	HF datalink
HSI	horizontal situation indicator
HUD	head-up display
Hz	Hertz (cycles per second)
IAF	initial approach fix
IAS	indicated air speed
IAWP	initial approach waypoint
ICAO	International Civil Aviation Organisation
i.c.w.	interrupted carrier wave (keying)
IDENT	identification function
IFF	identification friend or foe (early SSR)
IFR	instrument flight rules
ILS	instrument landing system
IM	inner marker
INMARSAT	international maritime satellite organisation
INS	inertial navigation system
IRS	inertial reference system
ITCZ	inter-tropical convergence zone (meteorological – see *Ground Studies for Pilots: Meterology*)
IWP	intermediate waypoint
JAA	European Joint Aviation Authorities
JAR	European Joint Airworthiness Regulations
JAR-OPS	JARs – Operations
kHz	kilohertz (10^3 Hertz)
km	kilometre(s)
kt	knot(s) – nautical mile(s) per hour
kts	knots
LAAS	local area augmentation service (GPS)
LAT	latitude
LCD	liquid crystal display

LCZ	localiser
LED	light emitting diode
LF	low frequency (30–300 kHz)
LI	lane identification
LLZ	localiser
LMM	locator middle marker (NDB)
L-NAV	lateral navigation (FMS and RNAV)
LOC	localiser
LOM	locator outer marker (NDB)
LONG	longitude
LORAN	long range aid to navigation
L/R	left/right
LUHF	lowest usable high frequency
LUT	local user terminal (SATCOM)
m	metre(s)
°M	degrees magnetic
MAHP	missed approach holding point
MAHWP	missed approach holding waypoint
MAP	missed approach point
MAWP	missed approach waypoint
MDA	minimum descent altitude
MDH	minimum descent height
MF	medium frequency (300–3000 kHz)
MHz	megahertz (10^6 Hertz)
MIC	microphone
min	minute(s) *or* minimum
MKR	marker
MLS	microwave landing system
MM	middle marker
MN	magnetic north
MP	multipulse
mph	statute miles per hour
m/s	metres per second
MTI	moving target indicator
MUF	maximum usable frequency
°N	degrees north of the equator
NAV	navigation (often used for VOR function)
NCU	navigation computer unit
NDB	non-directional beacon
nm	nautical mile
NOTAM	Notice to Airmen (aeronautical information signal)

OBI	omni-bearing indicator
OBS	omni-bearing selector
OM	outer marker
P	precision (as in guidance code of GPS)
°p	degrees port
PANS-OPS	Procedures for Air Navigation Services – Operations (ICAO Document 8168)
PAR	precision approach radar
PFD	primary flight display
PLB	personal locator beacon
POS	position
PPI	plan position indicator
PPS	precise positioning service (GPS)
pps	pulses per second
P-RAIM	predicted remote autonomous integrity monitor (GNSS)
prf	pulse repetition frequency
pri	pulse repetition interval
PRN	pseudo-random noise number (GPS satellite identification)
P-RNAV	precision area navigation system
prp	pulse repetition period
prr	pulse recurrence rate
PSR	primary search radar
QDL	(request for) series of magnetic bearings to the DF station
QDM	(request for) magnetic bearing to the DF station
QDR	(request for) magnetic bearing from the DF station
QE	quadrantal error (ADF)
QFE	(request for) atmospheric pressure at the aerodrome
QFF	meteorological office calculated sea level pressure at a reporting station
QGH	(request for) DF approach procedure
QNH	sea level pressure which will make an altimeter read aerodrome elevation on the airfield
QTE	(request for) true bearing from the DF station
QTF	(request for) position fix using DF
QUJ	(request for) true bearing to the DF station
°R	degrees relative
RA	resolution advisory (ACAS) *or* radio altitude
RAD	radar
Radalt	radio altimeter
RAIM	remote autonomous integrity monitor (GNSS)
RBI	relative bearing indicator
rel	relative

rel brg	relative bearing
RF	radio frequency
RMI	radio magnetic indicator
RNAV	area navigation system
RoC	rate of climb
RoD	rate of descent
RP	reference point *or* reporting point
rpm	revolutions per minute
RSR	en-route surveillance radar
R/T	radio telephony
RTF	radio telephony
RVR	runway visual range
RVSM	reduced vertical separation minimum
RW	runway
RWY	runway
Rx	receiver
s	second(s)
°S	degrees south of the equator *or* degrees starboard
SA	selective availability (GPS)
SAR	search and rescue
SARSAT	search and rescue satellite system (US)
SATCOM	satellite communications
SATNAV	satellite navigation
SBY	standby
SELCAL	selective calling system
SHF	super high frequency (3–30 GHz)
SID	standard instrument departure
Sig	signal
SMR	surface movement radar
SPI	special position indicator (SSR indent)
SPKR	loudspeaker
SPS	standard positioning service (GPS)
SRA	surveillance radar approach
SSB	single sideband
SSR	secondary surveillance radar
STAR	standard terminal arrival route
Stb	starboard
STBY	standby
°T	degrees true
TA	traffic advisory (ACAS)
TAD	terrain awareness and alerting display (TAWS)
TACAN	tactical air navigation system (military) using DME for range measurement

TAR	terminal area surveillance radar
TAS	true air speed
TAWS	terrain awareness and warning system
TCAS	traffic conflict alert system/traffic collision avoidance system (commercial)
TKE	track error
TMG	track made good
TR	track
TRSB	time referenced scanning beam (MLS)
TST	test
TV	television
TVOR	terminal VOR
Tx	transmitter
UHF	ultra high frequency (300–3000 MHz)
UK	the United Kingdom
UKAIP	the United Kingdom Aeronautical Information Publication
UTC	universal time constant (similar to Greenwich Mean Time but more precise)
Var	magnetic variation
VDF	VHF direction finding
VFR	visual flight rules
VHF	very high frequency (30–300 MHz)
VLF	very low frequency (3–30 kHz)
V-NAV	vertical navigation (FMS and RNAV)
VOR	VHF omni-directional range
VORTAC	co-located VOR and TACAN station
VOT	a test VOR station
VSI	vertical speed indicator
°W	degrees west of the prime meridian
WAAS	wide area augmentation service (DGPS)
WDGS	windshear detection and guidance system
WPT	waypoint
w/t	wireless telegraphy
WV	wind velocity
Wx	weather
Z	en-route Z marker
ρ (rho)	used to describe a circular position line as from a DME or GPS
λ (lamda)	wavelength
μ (mu)	micro – one millionth of a unit (as $1\,\mu s = 1$ microsecond)
θ (theta)	used to describe a radial position line as from a VOR

Chapter 1
Basic Radio

Introduction

In the early days of aviation, pilots used their eyes and a set of rules to find their way and to avoid obstacles and other aircraft. Such is the usual aim of most private pilots today. However, the weather is not always kind enough to allow those simple pleasures, and soon it became essential to communicate with other aircraft which might be hidden behind cloud, and to be able to navigate when out of sight of the ground.

Fortunately, the First World War encouraged not only the development of aircraft, but also of radio, the effects of which had been predicted by scientists in London even before Heinrich Hertz conducted his famous experiment in 1887 and proved the existence of electromagnetic waves. Hertz's experiments also showed the important facts that electromagnetic waves propagate in a vacuum and can be stopped by a metallic screen. He calculated their speed, and determined the relationship between frequency and wavelength.

Nowadays, radio waves are used in aviation not only for communication and basic navigation, but to a great extent as a means to ensure the safety of flight in a crowded sky by directing aircraft to follow an exact flight path in three dimensions, so that others can be kept at a safe distance. This book sets out to explain how the theory of electromagnetic waves is applied to achieve all the desired aims.

Simplified transmission and reception of radio waves

Radio waves are the product of the changing fields produced by an alternating current. Alternating current is produced by rotating a wire in a magnetic field (or by rotating the magnetic field itself). This makes electrons flow along the wire in accordance with the alternating voltage produced. Because the voltage and current are alternating, the electrons flow in one direction for half the rotation, and in the reverse direction for the second half. This electron flow, alternately forwards and backwards, means that the current is continually changing. This changing current in turn produces fields along the wire.

If the wire is a closed circuit, then the fields in one part of the wire are generally cancelled out by those in another part. However, an alternating current can be induced in an open circuit, with a bare wire at the end. In this case, the fields will propagate outwards from the wire in a normal (at 90° to it) direction. If the wire is the correct length, the fields will resonate and send continuous alternating waves of energy outwards. This outward propagation of the fields forms the transmitted radio waves.

If a wire of the same length is placed in the same direction in space as the transmitting aerial, the fields will affect the wire and induce an alternating current in it, so a receiver a considerable distance away can receive the transmitted signal exactly. This was basically what Marconi achieved in his experiments, and although modern technology uses sophisticated electronic devices, the effects are the same. In fact, the traditional aerial is called a half-wave dipole (see Figure 1.1), and is actually two halves, each the length of one quarter wavelength, fed from the middle.

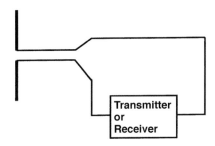

Figure 1.1   A half-wave dipole aerial.

The electromagnetic waves travel at constant speed through the vacuum of space. The speed varies depending on the density of whatever medium they are travelling through, and we shall see later how this affects the actual propagation around the earth, but in general the gases in our atmosphere change the speed by only a small amount. The speed of propagation of electromagnetic waves (often called the speed of light; light waves are electromagnetic waves) in air averages approximately 300 million metres per second (300×10^6 m/s).

The radio wave

As the waves are alternating fields, the terminology associated with alternating currents can be used in a similar fashion.

An a.c. voltage in a wire reverses its direction a number of times every

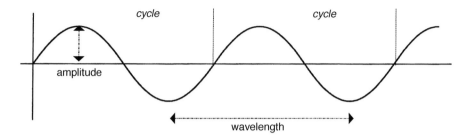

Figure 1.2   Graph representing a radio wave.

second. Consequently, if a graph of the current in the wire is plotted against time, it will be found to be in the form of a sine curve, as in Figure 1.2.

Cycle.   A cycle is one complete series of values, or one complete process.

Hertz.   One hertz is one cycle per second. The number of cycles per second is expressed in hertz.

Amplitude.   The amplitude of a curve is the maximum displacement, or the maximum value it achieves during a cycle. The amplitude of a radio wave is the maximum strength of the signal during a cycle. In any alternating case, the positive amplitude is the same as the negative amplitude. (The part of the curve in Figure 1.2 above the mean or time axis is called positive and the part below is called negative).

Frequency (f).   Frequency of an alternating current or a radio wave is the number of cycles occurring in one second, expressed in hertz (Hz). For example, 500 Hz is 500 cycles per second. Since the number of cycles per second of normal radio waves is very high it is usual to refer to their frequency in terms of kilohertz, megahertz and gigahertz as follows:

1 cycle per second	= 1 Hz (hertz)
1000 Hz	= 1 kHz (kilohertz)
1000 kHz	= 1 MHz (megahertz)
1000 MHz	= 1 GHz (gigahertz)

Wavelength (λ).   This is the physical distance travelled by the radio wave during one complete cycle of transmission. It is defined as the distance between successive crests or the distance between two consecutive points at which the moving particles of the medium have the same displacement from the mean value and are moving in the same direction.

Wavelength/frequency relationship

In one second, a radio wave will cover a geographical distance of 300 million metres. If the wave has a frequency of one hertz, the cycle will take one second to pass a point, and the geographical distance between the start of the cycle and its end will be 300 million metres. (When the end of the cycle reaches the point, the start of the cycle will be 300 million metres ahead.) That means that the length of the wave, or wavelength, is 300 million metres. If the frequency of the wave is two hertz, there will be two cycles passing the point in one second, and the wavelength will be half of 300 million metres, or 150 million metres. Thus as frequency is increased, the wavelength is decreased in the same proportion and vice versa; putting this in a formula:

$$\text{wavelength} = \frac{\text{speed of radio waves}}{\text{frequency}} \quad \text{or} \quad \lambda = \frac{c}{f}$$

and

$$\text{frequency} = \frac{\text{speed of radio waves}}{\text{wavelength}} \quad \text{or} \quad f = \frac{c}{\lambda}$$

By using the above formulae it is possible to convert frequency into wavelength and wavelength into frequency. To avoid any errors, at least at the beginning, basic units should be used in the formulae. The use of hertz for frequency gives metres for wavelength, and using metres for wavelength gives hertz for frequency which may then be expressed as kHz or MHz as appropriate for the answer.

Examples

(1)   If the wavelength is 1.5 km, what is the frequency?

$$\text{Frequency in Hz} = \frac{\text{speed in m/s}}{\text{wavelength in m}}$$

$$= \frac{300\,000\,000}{1500}$$

$$= 200\,000\,\text{Hz}$$

$$= 200\,\text{kHz}$$

(2)   If the transmission frequency is 75 MHz, what is the wavelength?

$$\text{Wavelength in metres} = \frac{\text{speed in m/s}}{\text{frequency in Hz}}$$

$$= \frac{300\,000\,000}{75\,000\,000}$$

$$= 4\,\text{m}$$

(3)   If the wavelength is 3 cm, what is the frequency?

$$3\,\text{cm} = 0.03\,\text{m}$$

$$\text{Frequency} = \frac{300\,000\,000}{0.03}$$

$$= 10\,000\,000\,000\,\text{Hz}$$

$$= 10\,\text{GHz}$$

(4)   If the frequency is 13 500 MHz, what is the wavelength?

$$\text{Wavelength} = \frac{300\,000\,000}{13\,500\,000\,000}$$

$$= \frac{3}{135}$$

$$= 0.0222\,\text{m}$$

$$= 2.22\,\text{cm}$$

(5)   How many wavelengths, to the nearest whole number, of frequency 150 MHz, are equivalent to 52 ft?

$$\text{Wavelength} = \frac{300\,000\,000}{150\,000\,000}$$

$$= 2\,\text{m}$$

$$= 2 \times 3.28\,\text{ft}$$

$$= 6.56\,\text{ft}$$

The number of times 6.56 will go into 52 ft $= \dfrac{52}{6.56}$

$$= 8\ (\text{approx})$$

Now try these; the answers are at the back of the book.

1. Wavelength is 3 m, what is the frequency?
2. Frequency is 100 kHz. What is the wavelength in metres?
3. Wavelength is 3520 m. What is the frequency?
4. Frequency 325 kHz, what is the wavelength?

5. Frequency 117 000 kHz, what is the wavelength?
6. Wavelength 3.41 cm, what is the frequency?
7. How many wavelengths to the nearest whole number is equivalent to 60 ft if the transmission frequency is 100 MHz? (1 m = 3.28 ft)
8. If wavelength is 2.739 m, what is the frequency?
9. Give the frequency appropriate to a wavelength of 2222 m.
10. If the frequency is 1429 kHz, what is the wavelength?

Phase and phase difference

Looking at Figure 1.2, we can talk about the cycle starting when the curve crosses the time axis on the way up. The rotating part of the generator would then be in a certain position relative to the static part. The rotating part (rotor) would turn through a complete circle between that point and the corresponding point at the end of the cycle. The complete circle is 360°, so we can mark the time axis in degrees of rotor movement from its start position. After 360°, the relative positions would be the same as at the beginning, so the time axis could be marked in repeating 360° cycles. We could refer to any position along the wave cycle as being at a certain number of degrees from the start. Because every cycle of the wave is the same, the amplitude of the wave at a certain angular position would be the same for every cycle.

We can compare two waves with the same frequency (and therefore wavelength), by their amplitudes. We can also compare them by looking at their cycle starting positions. If one of them reaches the start position 90° after the first, we say that it is 90° out of phase with the first. The maximum amplitude of the second wave will reach a particular point a quarter of a wavelength later than that of the first wave. We say that we are comparing the phase of the two waves. The actual angle of phase difference can be used for comparison, as in Figures 1.3(a), (b) and (c). This forms the principle of many radio navigation aids. It must be noted that this comparison can only be made between two waves of the same frequency.

Polarisation

As mentioned earlier, when a suitable a.c. is applied to an aerial, electromagnetic waves are radiated from the aerial. These waves alternate with the same frequency as the a.c. current applied to the aerial. The two components, electric and magnetic, thus radiated travel together at the speed of light. Both travel at right angles to each other, as illustrated in Figure 1.4, and also at right angles to the direction of propagation (the direction of propagation is away from the antenna).

When the transmission is being made from a vertical antenna, the electrical component, E, travels in the vertical plane, and its associated magnetic

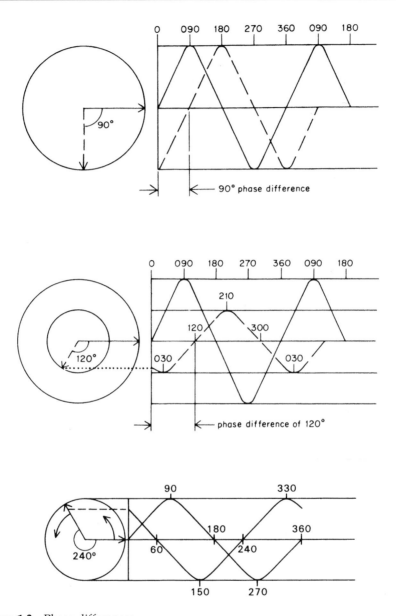

Figure 1.3   Phase differences.

component, *H*, in the horizontal plane. The emission is called vertically polarised. Similarly, for a horizontal aerial the electrical component travels in the horizontal plane, the magnetic component in the vertical plane and the emission is horizontally polarised. Where the electrical and magnetic components spin around the axis of advance, the signal is circularly polarised. (This technique is used in reducing rain clutter in radar.)

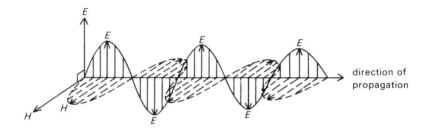

direction of
propagation

Figure 1.4   Radio wave polarisation.

The importance in knowing the polarisation lies in the orientation of the receiver aerial. A vertical aerial will efficiently receive the electrical component of a vertically polarised signal. If the receiver aerial, on the other hand, was perfectly horizontal, it would receive no electrical component. Similarly, a horizontal aerial will efficiently receive a horizontally polarised signal.

The lines joining the curve to the time axis (vector lengths) in Figure 1.4 represent the strength or field intensity of the signal at a given instant. As the signal travels further the energy spreads out in an ever increasing volume of space. This is one form of attenuation (reduction in signal strength) of the signals – attenuation due to spreadout. The reduction in signal strength is governed by the inverse square law in experimental conditions in a vacuum. Thus, if the field strength at a point at a given distance from the transmitter measures, say 80 microvolts, then the reading at another point twice the distance from the transmitter will be a quarter of the value, that is 20 microvolts.

Polar diagrams

The transmitted waves from a simple system with a single wire aerial travel in all directions around the antenna. This broadcasting means that the energy of the field is being attenuated as described above. Much of radio engineering is concerned with increasing the strength of the signal by amplifying it and sending it in particular directions, so that the strength of the signal is reduced as little as possible.

Radio engineers use what they call polar diagrams to show what is happening to the signal after it leaves the transmitting antenna. These show the position of points around the antenna where the strength of the signal has reduced to a certain level, often half its original strength. They are generally drawn as maps of the horizontal area around the aerial (horizontal polar diagrams). For a broadcast transmitter, the horizontal polar diagram is a circle, as in Figure 1.5 below, but for most navigation aids the polar diagram is designed to have a specific shape, such as in Figure 1.6.

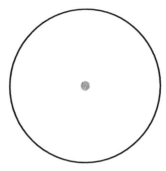

Figure 1.5   Circular polar diagram.

The strength of transmission in a particular direction is shown by the length of a vector, a line which can be drawn from the transmitter aerial to the edge of the polar diagram. In Figure 1.5 , the vector would be a radius of the circle. In Figure 1.6, vector OA represents the transmission strength in the direction of X. The polar diagram of a receiver (Rx) aerial similarly gives indication of reception from various directions. An aerial with a circular polar diagram would receive signals equally from all directions whereas a receiver aerial with a specially shaped polar diagram can be used to determine the direction of incoming signals. In Figure 1.6, if the receiving aerial was at O, it would receive maximum signals from a transmitter (Tx) in the direction of Y, and reduced signals from direction X.

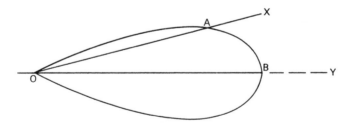

Figure 1.6   Directional polar diagram.

Aerials

The simple half-wave dipole was cumbersome for aircraft use. It was quickly found that placing a sheet of metal, such as the aircraft skin, below one half of the dipole, allowed engineers to dispense with the other half and still retain the characteristics of the dipole. This 'quarter-wave' antenna was for a long time the standard broadcast transmitting and receiving aerial on aircraft, and is still used.

Scientists found by experimentation that they could alter the characteristics of antennas by placing objects near them. To produce more power for transmission or reception in a particular direction (a directional polar diagram), modifications were made to the aerial design, as detailed below. A breakthrough in radio technology came when engineers were able to change the characteristics of short lengths of conducting material by electronic means. Careful shaping and specialised circuits were developed to allow short lengths of metal to simulate the characters of much longer ones.

Parasitic antennas

It was found that placing a length of metal close to a half-wave antenna affected the polar diagram. A rod of length slightly shorter than the antenna, placed slightly less than a quarter wavelength away, would strengthen the signal in its direction, effectively 'directing' the waves towards itself. Further 'parasitic radiators' or specifically 'directors' positioned equally spaced in the same direction, increase the signal strength in that direction. Similarly, a slightly longer rod placed slightly more than a quarter wavelength away strengthened the signal in the opposite direction, effectively 'reflecting' the signal. A combination of parasitic antennas, one reflector and several directors, called a 'yagi' aerial, constitutes a typical terrestrial television aerial, similar to that shown in Figure 1.7. For some time, these have had limited use in the field of aviation, although certain directional ground radars employed them instead of using parabolic reflectors. They have, however, become more common in ground equipments recently.

Figure 1.7   Directional yagi aerial.

Parabolic reflectors

Single pole reflectors produce an intensification of the signal strength in the opposite direction, but the polar diagram is only altered slightly (such an alteration is shown in Chapter 4 at Figure 4.1(c)). To produce a more directional signal, a full yagi aerial is required, but it is possible to use a group of reflectors arranged in the shape of a parabola to focus the beam in a particular direction. In fact, the parabolic reflector is more commonly used to focus the signal which has been transmitted to it along a 'wave guide'. This wave guide is a metal tube of rectangular cross-section, approximately

0.7 × 0.4 wavelengths, which allows a signal to pass along it with the minimum attenuation (loss of signal strength). A typical parabolic antenna is shown in Chapter 9, at Figure 9.1.

Modulation

A page left blank in a newspaper conveys no information. To provide information some writing or picture must be printed on it. A plain radio wave may be likened to a blank newspaper, it cannot be heard nor can it convey information. Special components can make it audible, but the only signal heard is a constant audio tone; still nothing is 'read'. Some form of intelligence must be 'impressed' upon such a wave if it is to convey information. The process of impressing such information by changing the original signal is called 'modulation', and it is done in a variety of ways. In all cases the radio waves simply act as a vehicle for the information, so they are commonly called 'carrier waves'. The waveform of information which is being impressed upon the carrier wave is called the 'modulating wave'. Some of the ways in which the carrier may be changed to transmit information are given below.

Keying

This is often called wireless telegraphy (w/t). It consists of starting and stopping the continuous carrier wave, breaking it up into dots and dashes, and so is sometimes also described as 'interrupted carrier wave' or i.c.w. The communication is by a code; groups of these dots and dashes being assigned particular meanings, as in Morse code. The technique is primarily used for long-distance communication; however, some radio navigation facilities may break their carriers for identification by dots and dashes. The receiver requires a beat frequency oscillator (BFO) facility, which is described in Chapter 5, to make the signals audible.

Amplitude modulation (AM)

This can be used either to radiate speech, music, etc., or to transmit coded messages at audio frequencies. As the name suggests, the amplitude (strength) of the carrier is varied in accordance with the amplitude of the audio modulating signal, keeping the carrier's frequency constant. In Figure 1.8, a modulating audio signal is impressed on a radio frequency, showing the resultant variation in amplitude of the carrier wave. In the third picture, lines have been drawn to represent the boundary of the modulated signal – the amplitude of the carrier wave fluctuates inside this 'modulation envelope'.

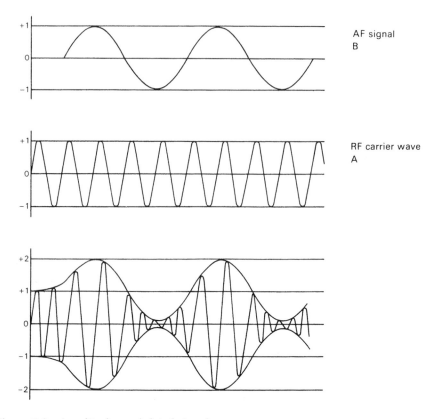

Figure 1.8   Amplitude modulated signal.

When a signal is amplitude modulated, its total amplitude varies between the sum of the amplitudes of the two signals and the difference between them. In Figure 1.8, the amplitude of the modulating signal is slightly less than the amplitude of the carrier wave, which is the ideal relationship for maximum signal strength. When we compare the two original amplitudes, we can say that there is a depth of modulation of almost 100%. If the modulation depth is low, the signal may be hidden by 'noise'. If the depth exceeds 100%, the boundaries of the modulation envelope will cross over, attempting to produce negative signals, and a badly distorted signal would be received. Readers may have heard a similar resultant howl when speakers talk too close to a microphone! It is better to avoid any such distortion by aiming for a depth slightly below 100%.

Sidebands

We have stated that amplitude modulation does not change the frequency of the carrier wave. However, adding or subtracting a signal to the carrier does in fact produce signals at slightly different frequencies, those at the sum and difference of the two waves. These signals at the new frequencies, when added together, are of the same strength as the modulating wave. If we measure frequencies, we can see that a simple amplitude modulated signal actually sends three waves; one at the original frequency, one at the sum of the two frequencies, and one at the difference between the two frequencies. Each of the two outside frequencies or sidebands is at half the original strength. Sidebands can be used to improve frequency band use, as seen below.

Simple transmitter

Figure 1.9 below shows a very much simplified diagram of a basic AM radio transmitter. The components shown are described below, together with additional components found in practical systems.

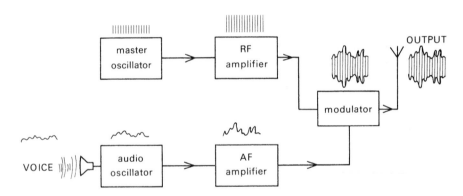

Figure 1.9   Simple AM transmitter.

Oscillator

The radio frequency carrier wave is generated in the oscillator, whose frequency may be controlled by one or a combination of several quartz crystals, a magnetron valve (now seldom used), or a semiconductor circuit incorporating 'varactor (variable capacitor) diodes'. These oscillations are usually at quite a low frequency. For anything above these low frequencies, one or more 'frequency multiplier' circuits must be included to bring the frequency of the oscillator up to that desired for transmission.

Radio frequency (RF) amplifier

Once the signal is at the correct frequency, it must be amplified until it is strong enough to pass through the remainder of the transmitter components clearly. Such amplifiers, like the other components, are now almost invariably semiconductor circuits.

Microphone and audio frequency (AF) amplifier

The audio frequency signal may be produced by the operator's microphone and/or an audio oscillator such as a recording device. That again must be amplified. Once amplified, any speech part of an audio signal will be processed in a 'speech processor' before modulation takes place. First the extreme frequencies (which make little difference to the receiver's understanding) are removed, and secondly, the amplitude variations in the speech are reduced to make the quieter parts easier to hear. This reduces the 'bandwidth' between the two sidebands.

Modulator

Here the amplified audio signals modulate the carrier wave as described above. The modulated signal is still not strong enough for transmission, so a further 'power amplifier', or series of amplifiers, is used to produce the required signal strength for transmission by the aerial.

Frequency modulation (FM)

Another method of modulating the carrier wave is to vary its frequency. The modulation involves changing the carrier frequency at a rate corresponding to the modulating signal's frequency, and at a frequency difference corresponding to the modulating frequency's amplitude. Modulation of a 1 MHz carrier by a simple 1 kHz sine wave signal would therefore produce a carrier which changed its frequency every 1000 waves corresponding to the amplitude of the 1 kHz signal at that time. The amplitude of the carrier is unaffected by the modulation. Figure 1.10 simulates an FM signal.

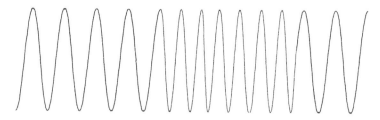

Figure 1.10 Frequency modulated signal.

Sidebands

A frequency modulated signal, by its very structure, transmits signals at varying frequencies. These appear in a band extending in relation to the strength of the modulating signal, around the central carrier frequency, and are distributed equally on both sides of it. Since all information in an FM signal is contained in these sidebands, the receiver must be able to pick up all of them. The range of frequencies containing the sidebands is called the bandwidth. Because all the information from the modulating signal is contained in mirrored sidebands around the carrier wave, as we shall see later it is possible to transmit and receive only one group of the sidebands. The other sideband, and even the carrier wave, can be suppressed and electronically replaced in the receiver. If a transmitter uses only the upper sideband, and receivers are tuned to receive that, another transmitter can use the lower sidebands for its signal. This effectively reduces the bandwidth required, and allows more transmitters to use a busy frequency band.

Comparing frequency and amplitude modulation

Power

Power is required to modulate any signal. For a given power input, an unmodulated carrier wave will travel a certain distance before it becomes lost in background 'static' or 'noise' internally generated in a receiver. An FM signal requires some extra power, and will travel a shorter distance, but an AM signal with maximum modulation requires about 50% more power than the basic carrier wave.

Transmitters and receivers

An AM transmitter is a complicated piece of equipment, whereas its receiver can be quite simple. Older readers may remember simple crystal receiver sets before transistor radios spread music across the world. In contrast, an FM transmitter is relatively simpler (although still using the same basic components as in Figure 1.9) but a suitable receiver must be more complex.

Static interference

Static interference is caused by electrical disturbances in the atmosphere. The word 'noise' is usually reserved for interference from electrical components in transmitters and receivers. These disturbances occur over all frequencies, but are worse in the lower bands. They are much more of a problem to AM reception, because the interference is similar to an AM signal. Most static can be filtered out of an FM receiver.

Pulse modulation

Both AM and FM transmissions use continuous carriers. Certain uses of radio waves, for example basic radar, require the signal, or parts of it, to be sent in short bursts. This was initially done by keying short transmissions with long gaps, and such pulse modulation can be applied to basic AM or FM signals. The aim can also be achieved by giving short pulses of either amplitude or frequency modulation to a continuous carrier.

Phase modulation

It is possible to alter the carrier wave by changing the phase of the transmitted signal. This can be used efficiently for transmitting digital information, as in global positioning system (GPS) signals, where the phase is reversed every time the binary digit changes (Figure 1.11).

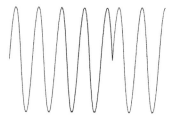

Figure 1.11   Phase modulation – reversing the phase.

Emission designation

Radio waves are often designated by a 3-symbol code, which is used by ICAO as part of the definitions of all aviation communication and navigation systems. For example, a VOR signal is designated as 'A9W'. The most important meanings of each part of the code are listed below. Several signals are described by more than one designator if the signal itself consists of more than one part. For example, a continuous wave signal with no modulation which for a short period is modulated in amplitude using a single modulating sub-carrier for identification purposes may be designated as N0N A2A.

First symbol

This tells the type of modulation on the main carrier wave. This includes:

N   No modulation
A   Amplitude modulated, double sideband
H   Amplitude modulated, single sideband and carrier wave
J   Amplitude modulated, single sideband, suppressed carrier wave
F   Frequency modulated
G   Phase modulated
P   Pulse modulated, constant amplitude
K   Pulse modulated, amplitude modulated

Second symbol

This designates the nature of the signal or signals modulating the main carrier:

0   No modulating symbol
1   Single channel containing quantised or digital information without the use of a modulating sub-carrier
2   Single channel containing quantised or digital information, using a modulating sub-carrier
3   Single channel containing analogue information
7   Two or more channels containing quantised or digital information
8   Two or more channels containing analogue information
9   Composite system comprising 1, 2 or 7 above, with 3 or 8 above
X   Cases not otherwise covered

Third symbol

Type of information transmitted. (This does not include information carried by the presence of the waves.)

N   No information transmitted
A   Telegraphy – for aural reception
B   Telegraphy – for automatic reception
C   Facsimile
D   Data transmission, telemetry, telecommand
E   Telephony (including sound broadcasting)
F   Television (video)
W   Combination of the above
X   Cases not otherwise covered

Summary

1. Radio waves travel at 300 million metres per second
2. Frequency $= \dfrac{\text{speed of light}}{\text{wavelength}}$ and wavelength $= \dfrac{\text{speed of light}}{\text{frequency}}$
3. The magnetic field propagates at 90° to the electrical field. The wave is polarised in the direction of the electrical field
4. AM signals can be modulated to a maximum depth of 100%
5. AM transmitters are more complex than FM transmitters
6. FM receivers are more complex than AM receivers
7. Static affects AM signals more

Sample questions

1. If a radio wave is described as 'horizontally polarised', in what planes are the (i) magnetic and (ii) electrical fields oriented?

   a. (i) horizontal   (ii) horizontal
   b. (i) horizontal   (ii) vertical
   c. (i) vertical   (ii) horizontal
   d. (i) horizontal   (ii) circular

2. Which emission designation would apply to a single channel amplitude modulated single sideband signal with a suppressed carrier wave carrying analogue information for aural (telephonic) reception?

   a. A2A
   b. A8W
   c. J3E
   d. P0N

3. Which of the following facilities is most likely to use a 'frequency modulated' signal?

   a. Ground–ground UHF radio telephony
   b. Air–air VHF radio telephony
   c. Ground–air HF data link
   d. MF radio navigation beacon

4. What is a hertz?

   a. the wavelength in cycles per second
   b. a frequency of one cycle per second
   c. the wavelength corresponding to one cycle per second
   d. the frequency in cycles per second

5. If wavelength is 8 mm, what is the frequency?

    a. 37.5 GHz
    b. 375 GHz
    c. 3750 GHz
    d. 3.75 GHz

6. For a frequency of 200 kHz, what is the wavelength?

    a. 1500 m
    b. 150 m
    c. 1500 km
    d. 150 km

7. In the diagram below, how far out of phase are the two radio waves represented?

    a. 45°
    b. 90°
    c. 180°
    d. 225°

8. An AM signal at frequency f_m carried on a transmitted frequency f_c produces:

    a. one sideband of transmission at $f_c + 2f_m$
    b. two sidebands of transmission at $(f_c + f_m)$ and $(f_c - f_m)$
    c. one sideband of transmission at $f_c - 2f_m$
    d. two sidebands of transmission at $(f_c + 2f_m)$ and $(f_c - 2f_m)$

Chapter 2
Radio Waves in the Atmosphere

Introduction

As explained in the introduction to Chapter 1, undisturbed radio waves in space travel in straight lines at a constant calculated speed. However, the earth is an uneven mass of solids and liquids, surrounded by a mixture of gases with varying density and electrical charge. All of these factors affect the 'propagation' of the waves to a greater or lesser effect. In this chapter we shall look at the different ways in which waves of different frequencies are affected by our earth and its atmosphere.

Radio spectrum

Before investigating the effects on propagation, we should look at the way in which the spectrum of electromagnetic waves used for radio is traditionally divided into bands. The whole electromagnetic spectrum includes radiation in the form of light, X-rays and gamma rays, but radio waves comprise only the bottom end of the complete spectrum. Voice frequencies fall immediately below this radio spectrum, but sound waves are actually pressure waves and are propagated differently from electromagnetic waves. The internationally recognised names for the frequency bands are given in Table 2.1 below. However, sometimes small groups of frequencies are regarded as part of a neighbouring band, and other designators are sometimes given to specific bands of frequencies.

Table 2.1 Frequency band designators.

Frequency band name	Abbreviation	Frequencies	Wavelengths
very low frequency	VLF	3–30 kHz	100–10 km
low frequency	LF	30–300 kHz	10 000–1000 m
medium frequency	MF	300–3000 kHz	1000–100 m
high frequency	HF	3–30 MHz	100–10 m
very high frequency	VHF	30–300 MHz	10–1 m
ultra high frequency	UHF	300–3000 MHz	100–10 cm
super high frequency	SHF	3–30 GHz	10–1 cm
extremely high frequency	EHF	30–300 GHz	100–10 mm

Some of the other common names for specific groups of frequencies within the UHF and SHF bands are listed below (Table 2.2), as they will be referred to later in this book.

Table 2.2 Radar frequency band designators.

Frequency band name	Frequencies
Radar L band	1–2 GHz
Radar S band	2–4 GHz
Radar C band	4–8 GHz
Radar X band	8–12.5 GHz

The surface of the earth

The earth is mainly solid rock, covered with liquid water over the majority of its surface, loam and sand over the low lying land areas, and ice close to the poles. Radio waves of VHF bands and above are generally reflected by solid material, although lower frequencies are absorbed. However, the surface of the earth also affects waves travelling parallel or almost parallel to it. At low frequencies, some of the wave energy is lost in inducing currents in the surface, which also slows the wave down. The amount lost depends on the material and its condition affecting its conductivity; dry sand produces greater attenuation and speed reduction than wet loam, and sea water produces less attenuation than either.

Obstacles in the path of a radio wave also affect its path. Radio waves tend to be reflected by objects larger than about half their wavelength. At higher frequencies, most obstacles will cause reflection, or absorption, and therefore shadows behind them, but at lower frequencies the waves will curve around a small obstacle, even a hill. This is called 'diffraction', and can be considered as the obstacle creating 'friction' in the part of the wave close to it, causing the wave to curve towards it as it passes. The amount of diffraction is inversely proportional to the frequency. Radio waves can also be redirected by scattering between molecules in the atmosphere, and reflected from neighbouring solid objects.

One would expect a radio wave transmitted from an aerial on the surface of the earth to travel in all directions at or above a tangent to the surface. Since an aerial will usually be positioned above the actual surface of the earth, the path of the signal closest to the earth's surface would actually be a tangent to the surface from the top of the antenna, as shown in Figure 2.1. The maximum range at which a signal could be received can be calculated mathematically. However, the attenuation and diffraction of the signal by the surface of the earth will cause a small amount of bending towards that

Figure 2.1   Ideal line of sight propagation.

surface. Even at higher frequencies, that is measurable, and coupled with the more important effects detailed in the next section means that we can find the actual maximum effective range D of a VHF or UHF signal from an aerial with altitude H_1 to another antenna with altitude H_2 using the formula

$$D = 1.25\sqrt{H_1} + 1.25\sqrt{H_2}$$

where D is in nautical miles, and H_1 and H_2 are in feet above mean sea level. At lower frequencies, diffraction and attenuation increase, and some of the signal will actually follow the surface of the earth.

The lower atmosphere

The molecules of gas in the atmosphere absorb some of the energy in the radio wave. This attenuation depends on the wavelength of the signal. The shorter the wavelength (or the higher the frequency) the greater the atmospheric absorption.

The molecules of gas in the air can also reflect some radio energy, especially in the UHF band and above. This 'troposcatter' effect can be used by sending a signal from a directional aerial towards a receiver, but slightly upwards. A directional receiver over the horizon can collect any scattered signal which continues in its direction. However, this is not usually employed for aviation purposes.

The density of air reduces with pressure, but increases with temperature. The radio signal travels faster in a less dense medium, and if a wave passes through gas of changing density at an angle, it will curve towards the higher density. Density normally reduces slowly with altitude in the troposphere, where the pressure reduction has more effect than the temperature reduction, and in the stratosphere an increasing temperature with altitude reduces the density further. This effect increases the bending of radio waves around the earth's surface, and can also produce more spectacular results.

The speed of radio waves also changes with the different gases. Water vapour is less dense than dry air, and changes in humidity suggest a similar bending towards less humid air. However, a high water vapour content actually encourages refraction (bending) towards it.

One of the more spectacular results is tropospheric ducting, also called 'super-refraction'. This is sometimes caused when high surface air pressure produces a temperature inversion above the earth's surface. Under the inversion, especially over the sea, the air can become warm and very moist. The sinking air above the inversion has become notably drier, and cooler than the air just below it. A radio signal can be refracted downwards quite steeply by this moist inversion, and can then be reflected on contact with the earth, refracted at the inversion again, and reflected again, allowing a VHF or UHF signal to propagate for long distances by 'bouncing' from top to bottom of the duct, as in Figure 2.2. Readers may have noticed the effect of this on their TV screens, when the reflected signal from a transmitter a long way away arrives out of phase with the stronger primary signal from the usual transmitter. This causes 'ghosting' on TV sets, and is often referred to by TV weather presenters when an anticyclone is present (see *Ground Studies for Pilots: Meteorology*). Ducting is most common over the sea in the tropical and subtropical areas. It can also occur in inversions at height.

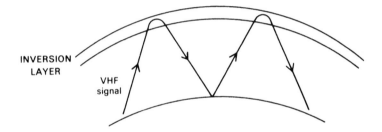

Figure 2.2   Super-refraction.

The opposite effect can take place if the relative densities are arranged in such a way that the signal is refracted upwards and away from the surface of the earth. This effectively raises the radio horizon above the normal line-of-sight, and reduces the range at which a signal can be received by a receiver close to the surface of the earth. This effect being opposite to 'super-refraction' is called 'sub-refraction'.

The troposphere is the region of the atmosphere where all the weather occurs. Moisture is spread upwards by convection or frontal lifting, and thunderstorms occur over most of the earth. Every thunderstorm produces electrical effects, including emissions of electromagnetic waves. This 'static' is mainly in the lower frequency bands, and is often strong enough to mask transmitted signals at long ranges.

The ionosphere

Above the tropopause lies the stratosphere, and above that a region called the ionosphere. Here radiation from the sun has a considerable effect on the molecules of a thin atmosphere, and electrons are set free from their atoms. The free electrons provide several electrically charged layers in this iono-sphere, but their existence depends on excitation from the sun's rays. The number of free electrons, and their distribution, depend on the angle at which the sun's rays meet the ionosphere, as well as the intensity of the rays themselves.

The layers were discovered by their effect on radio waves, and indeed they have a considerable effect. As the density of free electrons changes, it changes the 'refractive index' of the air. Electromagnetic waves passing through the layers in the ionosphere at an angle are refracted, or bent, away from areas of higher electron density, which happen to be in the higher part of the ionosphere. Therefore radio waves are bent towards the earth. The amount of refraction depends on three factors; the frequency of the waves, the change in electron density, and the angle at which the waves hit the layer. The waves are also attenuated, by an amount depending on the electron density and the frequency.

The ionosphere is traditionally divided into three refractive layers, the top one of which is usually subdivided into two. The accepted distribution is given in Figure 2.3, but the existence and heights of the various layers depend on the sun, so they change continually. In the darkness of the earth's shadow, there is no sunlight, so the lowest parts of the ionosphere lose their free electrons at night. At greater distances from the earth, the sun's rays are diffracted around it, and continue to have an effect on the higher parts of the ionosphere, so even at night there are effects from the higher layers, in fact it seems that the top layers actually move upwards.

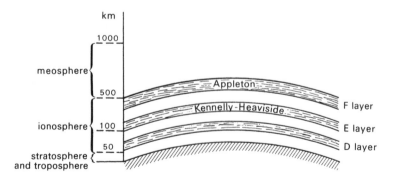

Figure 2.3   Ionosphere layers.

The D layer is generally regarded as being between 50 and 100 km above the surface of the earth, with an average altitude of 75 km. The E layer exists between 100 and 150 km, with an average altitude of 125 km. The F layer spreads between 150 and 350 km, with an average altitude of 225 km, although during the day it appears to split into a layer at the bottom of that (the F_1 layer), and a layer (F_2) in the upper part.

The D layer, where air density is high, and electron density is comparatively low, tends to absorb radio waves rather than refract them. The E layer, with greater electron density of up to $10^5/cm^3$ and less air density, produces some refraction of waves in the HF band, and the F layers with even lower air density and higher electron density (up to $10^6/cm^3$) do most of the refracting. Waves refracted at low levels will be refracted further at higher levels, provided they are not absorbed before then.

The refraction of electromagnetic waves in the ionosphere can be sufficient to bend a signal sent skyward down towards the earth again. We use this facility in HF communication, but it can cause problems when using MF navigation aids.

Diurnal effect

During the day, the sun's rays excite all parts of the ionosphere, and all layers are very active. The D layer absorbs a certain amount of power, especially at lower frequencies, and will also refract those low frequencies a certain amount. The E layer absorbs less but refracts more, and the F layers continue the effect.

At night, the D layer effectively disappears. The E layer thins, and appears to move upwards away from the earth. The F_1 layer thins also, and the area of strongest electron density appears to move upwards to join the F_2 layer, as shown in Figure 2.4 below. You can see that there are considerable changes between day and night, and these changes take place during twilight. Since the ionosphere tends to be used for long-range signals, twilight tends to vary between the transmission point and reception point, and reception at these

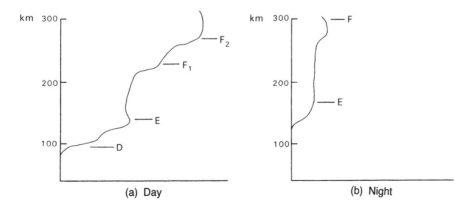

Figure 2.4 Ionosphere diurnal effects.

times tends to be difficult. These are critical periods for operating Automatic Direction Finding equipment (ADF).

Other effects

The tilt of the earth and its elliptical orbit all affect the amount of solar radiation received over different parts of the earth, and therefore the electron density of the ionosphere. The amount of radiation from the sun is also important, and there are large differences in ionisation at different stages of the 11-year sunspot cycle.

Ionospheric reflection

At very low frequencies and shallow angles of incidence to the D layer, it is possible for radio waves to be actually reflected from it, causing occasional interference at a different phase from the main signal.

Conditions of ionospheric refraction

Critical angle

The angle at which a radio signal strikes a layer is a major factor in deciding whether a signal will return to the surface of the earth or not. If it strikes the layer at a small angle to the perpendicular, it will not be refracted sufficiently to return. As the angle to the perpendicular progressively increases, the signal will bend progressively more, until at a critical angle, the signal will refract enough to return to the earth. This critical angle is measured from the perpendicular (a line normal to the earth's surface). The critical angle depends on the ionospheric conditions at the time. It also depends on the frequency of the signal; a lower frequency will bend more, and therefore have a lower critical angle. A frequency of more than 30 MHz will not usually return to earth.

Skip distance

At the critical angle, the signal returning to the earth will reach the surface at a point a certain distance away from the transmitter. A signal transmitted at more than the critical angle will return to earth at a greater range. A signal transmitted at less than the critical angle will not return, so the point at which the signal at the critical angle returns (the first sky wave) is the minimum range of the refracted wave. The distance between the transmitter and receiver in this case is called the 'skip distance', as seen in Figure 2.5.

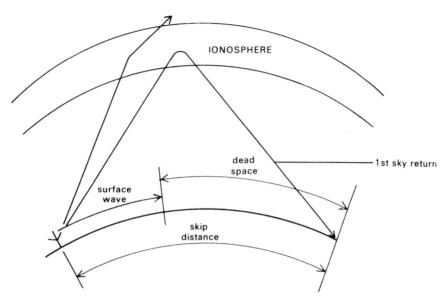

Figure 2.5   Critical angle and skip distance.

Interference

As we have seen, a wave transmitted from an aerial can get to a receiver along many paths. Although the signal will be attenuated by varying amounts along each path, often more than one receivable signal will arrive at the receiver. These signals will usually be out of phase, and if so will interfere with each other, causing 'fading'. If one signal is notably stronger than the other(s), the lower strength signals can be filtered out. However, if two paths allow signals of similar strengths to arrive at the receiver, this creates problems, described as multi-path errors. (Of course, if the signals arrive in phase, the signal strength will be increased.)

Interference is also caused by signals other than those from the same transmitter. Most frequencies are shared by transmitters in different geographical positions, because under normal circumstances their signals would not reach the receivers designed for the other transmitters. However, as we have seen, ionospheric or tropospheric conditions can send signals over much longer ranges than usual, and such interference can affect communication or, more dangerously, the accuracy of navigation aids.

Propagation types

To bring all the previous paragraphs together, we should look at the properties of each path of propagation, and how these affect emissions in the different frequency bands.

Direct waves

The waves reaching a receiver in a straight line (line-of-sight) are called direct waves. All frequencies can be received along direct waves. Signals are attenuated by spreadout in accordance with the inverse square law, such that the strength of the signal at a certain distance from the transmitter is four times as great as the signal at a point twice as far from the transmitter. In addition, as wavelengths reduce into the SHF and EHF bands, water drops and then the gas molecules in air can scatter and absorb progressively more of the signal.

Direct waves are regarded as the sole means of propagation of all signals in the VHF band and higher frequencies, and allow lower frequency signals to be received at short range.

Reflected waves

Waves can be reflected by any object whose size is more than half their wavelength. This is usually a hindrance to efficient propagation, but radar of course uses the principle of reflection to work. Direct waves and waves reflected from the ground are together called 'space waves'.

Surface waves

Surface waves are those which are bent around the surface of the earth. At HF frequencies or lower, the waves are refracted sufficiently to follow the curvature of the earth. However, there is considerable absorption by the earth's surface, and the higher the frequency the more absorption takes place. The range of a signal therefore is indirectly proportional to its frequency, or directly proportional to its wavelength, as well as being directly proportional to the power at the transmitter. Surface waves and space waves are together called 'ground waves'.

Surface waves are the primary means of propagation in the MF band, virtually the sole means of propagation in the LF band and lower frequencies, and the means of transmitting HF signals to receivers outside the range of direct waves but too close to receive sky waves.

Sky waves

The waves refracted by the ionosphere are called sky waves. The ionosphere absorbs and refracts signals by an amount directly proportional to their wavelength. At frequencies above the HF band, signals pass through the layers without returning, although they have been affected to some extent, quite considerably in the case of VHF signals. It is normally accepted that frequencies of 30 MHz and below will return to the earth in daylight, and at slightly lower frequencies also at night. As frequency reduces, the amount of refraction but also of attenuation increases, so at low frequencies a lot of transmitter power is required to send a sky wave which is strong enough to be received stronger than the static and noise.

If a sky wave is powerful enough, it can be reflected by the earth's surface and sent back into the ionosphere for a further refraction back to the earth. This is called 'multi-hop', and makes it possible for a sky wave signal to reach around the earth.

However, the minimum range of a sky wave is constrained by the skip distance at the critical angle. Since the critical angle (from the perpendicular) is directly proportional to the frequency, the minimum range is also proportional to the frequency for given ionospheric conditions. The higher the frequency, the further the minimum range, so the lower the frequency or longer the wavelength, the shorter the minimum range. Sky waves are the primary means of propagation of signals in the HF band, and can be used for signals in the MF band. However, at other frequencies, they are more a problem than a means of deliberate propagation.

Dead space

HF signals travel by both surface waves and sky waves. The maximum range of the surface waves depends on wavelength, and the minimum range also depends on wavelength. The gap between the two is called the 'dead space', as seen in Figure 2.5 above. The dead space reduces as the wavelength increases (frequency decreases).

Summary

Basic	Radio waves travel in straight lines at local speed of light (C)
	C varies slightly inverse to density
	Waves curve towards greater density
	Frequency bands – see Table 2.1
Direct waves	Line of sight straight lines (actually curve towards surface)
	Range (nm) = VHF formula $(1.25\sqrt{H_1} + 1.25\sqrt{H_2})$ (H in feet)
Space waves	Direct + ground reflected wave (possible phase differences)

Surface wave	Follows curve of the Earth
	Attenuation greatest at high frequencies over rough surfaces
	Range $\propto \sqrt{\text{power}}$
Ground waves	Space waves + surface waves
Sky waves	Refracted in ionosphere (varying electron density)
	Refraction greatest at low frequencies – short skip distances
	Absorption greatest at low frequencies – needs more power
Critical angle	Angle from vertical $-0°$ at 3 MHz, 90° at 30 MHz (approx)
	Critical angle = minimum angle at which waves return to Earth
Skip distance	Distance from transmitter to first returning sky wave
	Low frequency = short skip distance
Dead space	Skip distance minus surface wave range = dead space
	High frequency increases dead space from both ends
Ionosphere	D layer = 50–80 km, absorbs lots, refracts only LF
	E layer = 100–150 km, absorbs a fair bit, refracts MF and low HF
	F layer = 150–400 km, absorbs a little, refracts HF
Diurnal change	D layer disappears at night
	E layer thins, absorbs less, refracts at higher levels
	F layer thins, refracts at higher levels
Night	Given HF frequency is absorbed less, but greater critical angle
	For same distance use $\frac{1}{2}$ the frequency at night
Fading	Interference between phases of signals from different paths
Multi-hop	Powerful sky waves reflected back to ionosphere from Earth again
Duct propagation	VHF waves refracted in changing density in moist inversion

Sample questions

1. What happens to the direction and speed of propagation of radio waves as they pass through air whose density is increasing?

   a. The waves bend towards the high density and speed up
   b. The waves bend towards the high density and slow down

    c. The waves bend away from the high density and speed up

    d. The waves bend away from the high density and slow down

2. Which of the following is true?

    a. Ground reflected waves may interfere with surface waves

    b. Ground waves include space waves and surface waves

    c. Sky waves and direct waves combine to produce space waves

    d. Direct waves do not curve at all

3. Which of the following is true?

    a. Sky waves are refracted most at high frequencies

    b. Surface waves are attenuated most at low frequencies

    c. Sky waves are absorbed most at low frequencies

    d. The range of sky waves is proportional to the square root of the transmitted power

4. Which of the following is untrue about sky waves?

    a. The skip distance increases with frequency

    b. The dead space increases with frequency

    c. The critical angle increases with frequency

    d. Absorption increases with frequency

5. Which of the following is true about HF transmissions?

    a. The critical angle of a 30 MHz signal is about $10°$ during the day

    b. The skip distance of a given signal increases at night

    c. The range of the surface wave increases with frequency

    d. The dead space increases with frequency

6. Which of the following is true about the ionosphere?

    a. The D layer extends from approx 50 to 80 km and reflects LF signals

    b. The F_1 layer extends from approx 150 to 250 km and refracts VHF

    c. The E layer extends from approx 100 to 150 km and refracts MF

    d. The F_2 layer varies with season and time and refracts LF

7. Which of the following is true about the ionosphere?

    a. The density of charged particles is greatest in the D layer

    b. VHF signals are refracted back to earth in the F layer

    c. LF signals penetrate all the ionosphere layers

    d. The E layer thins and rises at night

8. What is a representative critical angle for a signal with a frequency of 3 MHz?

   a. 90° by day and night
   b. 90° by day, less at night
   c. 0° by day and night
   d. 0° by day, more at night

9. If you are using HF voice communications, and have good two-way signal during the day, what must you do as night falls to maintain communications?

   a. Nothing
   b. Change to a higher frequency
   c. Change to a lower frequency
   d. Stay at the same frequency but increase the volume

10. What happens in practical terms to the layers of the ionosphere at night?

    a. The D layer thickens and the E layer rises
    b. The D layer thins and the E layer rises
    c. The E layer thickens and the D layer rises
    d. The E layer thins and the D layer rises

11. If the transmission power of a signal sent by both surface wave and sky wave is doubled, what will happen to the dead space?

    a. Nothing
    b. The dead space will reduce from both ends
    c. The dead space will reduce as skip distance reduces
    d. The dead space will reduce from the transmission end

12. If the transmission frequency of a signal sent by both surface wave and sky wave is doubled, what will happen to the dead space?

    a. Nothing
    b. The dead space will reduce from both ends
    c. The dead space will reduce as skip distance reduces
    d. The dead space will reduce from the transmission end

13. If the transmission power of a signal sent by both surface wave and sky wave is doubled, what will happen to the range of the surface wave?

    a. Nothing
    b. It will double

c. It will quadruple

d. It will increase by a factor of 1.4

14. What happens to the maximum usable frequency of an HF signal between two points at night?

a. Nothing

b. It doubles

c. It halves

d. It reduces by a factor of 1.4

Chapter 3
Communications

Introduction

Communication in aviation is currently mainly achieved by voice modulation of radio waves. The future seems to lie in data transfer, which can be achieved without using the human voice. Nonetheless, voice communication is still important for the safe movement of air traffic, and will remain so in many parts of the world for some time.

Long-range communication – choice of frequency band

To achieve communication on the basis of global distances, the choice traditionally lay in the bands between VLF and HF, the frequency bands above HF being limited to direct wave, or 'line-of-sight' propagation. Although these higher frequency bands can now be used in association with satellite technology, many parts of the world still require this traditional means of communication.

Starting at the lowest end, we could obtain very long ranges in the VLF and LF bands and settle for them without further ado, but there are some inherent disadvantages in the employment of these bands. Just two requirements, of aerial size and power alone, are sufficiently forbidding to spur researchers to investigate alternative possibilities.

These possibilities are MF and HF. Of these two, HF is considered to be far superior:

- aerials are shorter and less expensive to install
- static noise is less than in MF and tolerable
- by using sky waves day and night, very long ranges are obtained for relatively less power
- higher frequencies suffer less attenuation in the ionosphere
- efficiency is further increased by beaming the radiation in the direction of the receiver

HF communications

The principle of efficient HF communication relies on choosing a frequency appropriate for a given set of ionospheric conditions that will produce the first return at the required skip distance from the transmitter. If the height of the refracting layer is known, the signal's path from the transmitter to the receiver via the ionosphere can be plotted and, from this, the angle of incidence the signal makes at the ionosphere can be calculated. An operator can use the angle of incidence to find the frequency whose critical angle that equates to. That frequency is the maximum usable frequency which will give communication at the estimated range, given the prevailing ionospheric conditions.

If we use a frequency higher than this maximum usable frequency, the signal will return beyond the receiver. At the maximum usable frequency itself, any ionospheric disturbance may increase the skip distance and cause the signal to be lost, so a slightly lower frequency is used. As we lower the frequency, attenuation increases and we need more transmitter power to produce an acceptable signal, until we are unable to produce enough power. When this limit is reached, we have reached the minimum usable frequency (LUHF or lowest usable high frequency).

In practice, graphs and nomograms are made available to the radio stations from which these values are directly extracted. The graphs take into consideration such factors as the station's position in latitude and longitude, time of the day, density of the ionosphere and any abnormal condition prevailing, and the distance at which the first sky return is required. Nowadays, of course, computers make the calculations, and can automatically select the optimum frequency for communication between the aircraft and any required ground station.

Because of the diurnal variation in the ionospheric density (see Chapter 2), if transmission is continued at night on a daytime frequency, a longer skip distance will result, leaving the receiver in the 'dead space'. This is because at night, as we have seen, the electron density decreases; the signal travels higher in the ionosphere before refraction, and is refracted less. For these reasons, the working frequency is lowered at night. This lowering of the frequency adjusts the skip distance because the lower frequencies are refracted more, and at lower levels. Attenuation is also less, despite the lower frequency, because the electron density is less.

The HF frequency band allocated to commercial aviation ranges from 2 MHz to 22 MHz, but in practice it is only used up to around 18 MHz. The Aeronautical Information Publication (AIP) for each country lists each Air Traffic Control Centre (ATCC) or Area Control Centre (ACC) ground station with the frequencies available which aircraft can use to communicate with them. The transmissions are amplitude modulated and a single sideband (SSB) emission, coded J3E, is used to economise on power and bandwidth or channel space.

In the early days when MF and HF w/t was in the forefront, aircraft were equipped with a trailing aerial. It consisted of a coil of wire which was wound out and held downwards by a weight. Normally it disappeared at the first sight of a thunderstorm, either by the pilot for safety or in the turbulence. In another system, a permanently fixed wire was used, stretching along the length of the fuselage. These aerials have now been replaced by recessed aerials electronically adjusted and conveniently located to give all-round reception to ground stations. To give an indication of power required, a mere 100 W transmitter can provide transatlantic voice communication.

Factors affecting HF range

- Transmission power
- Time of day; this affects the electron density
- Season of the year; this affects the electron density
- Any disturbances in the ionosphere (solar flares, etc.)
- Geographical location
- Frequency in use; this determines the critical angle and the depth of ionospheric penetration.

HF Datalink

HF Datalink (HFDL) is a facility used in Oceanic Control to send and receive information over normal HF frequencies, using the upper sideband of the selected frequency. The signal is phase modulated to send digital information. Modern equipment converts voice signals into similar digital information (like a digital mobile telephone), and vice versa, to provide digital voice communications.

The advantages claimed for digital HF, whether data or voice, include more rapid initial establishment of the communications link because of the automatic frequency selection. Once established, the link can be maintained continuously without a crew member constantly having to make transmissions, which allows messages to be passed quickly. A major benefit is that voice signal clarity is greatly improved by converting the message into digital form.

With the advent of satellite communications (see below), HF is losing its importance in oceanic flight. However, routes over and close to the North Pole, which are outside the cover of geostationary satellites, are becoming more common. HF communications, by voice and datalink, are likely to remain vital in such areas. Communications computers can control all the radios in an aircraft, and while receiving signals from all of them, can select automatically the most useful method of sending whatever signal the crew or the aircraft flight management computer wishes to send.

Short-range communication – choice of frequency band

There is a requirement to provide communication out to 80 nm range at 5000 ft, and 200 nm at 20 000 ft. As these are very short ranges, frequency bands from VLF to HF, with their disadvantages of complexity and static interference, are not necessary. The VHF band provides a practical facility.

At frequencies above VHF, aerial requirements become more complicated. The signal strength received by a simple antenna at a given range is proportional to the wavelength. Thus a longer wavelength (lower frequency) will give better reception.

VHF communication

The VHF band is chosen for RTF communication at short ranges, the operating frequencies being kept at the lower end of the band, 117.975 MHz to 137.000 MHz. Within this band communications channels are available at 8 kHz (actually 8.33 recurring) separation, although older equipments are still available at 25 kHz or even 50 kHz separation. The transmission is amplitude modulated, the type of emission being A3E. A transmitter producing 20 W power would be considered adequate for the intended ranges.

VHF is practically free from static, but being vertically polarised the receiver aerials do pick up some background noise. If absolute clarity of reception were required, a frequency modulated UHF signal could provide that, but the equipment would become more complex and expensive.

Frequency allocation

The highest frequencies in the band, from 136.900 to 136.975 MHz, are reserved for datalink purposes. Originally, VHF frequencies were allocated at 100 kHz spacing. The spacing progressively reduced through 50 kHz, to 25 kHz, and finally, by the time of writing, to 8.3 recurring kHz. Older radios were kept in service as the frequency spacing reduced, and most ATS frequencies outside controlled airspace were still allocated at 50 kHz spacing at the time of writing. Frequencies in controlled airspace were allocated at 25 kHz spacing, but mainly at high levels.

Figure 3.1 shows a typical VHF 25 kHz spacing communications control unit. The frequency marked 'use' is the frequency which the pilot is using for communication at present. The frequency marked 'stby' can be instantly selected by pressing the button on the bottom. The double knob on the right changes the frequency in the 'stby' window, and pulling the knob out increases the frequency by 25 kHz. The small knob marker 'off' and 'pull test' is the volume control. The unit shown has the facility to memorise

Figure 3.1   VHF communications control unit (*courtesy of Honeywell*).

particular frequencies and allocate channel numbers to them. The channel button switches the function of the main control between frequency selection and channel number selection.

The newer 8.33 kHz frequency spacing was introduced in 2000 in the most congested airspace in Europe, above Flight Level (FL) 245. It was not possible to make old radio receivers compatible with the new frequency spacing, because they had been designed with broad bandwidths to accept signals 7.5 kHz removed from the basic frequencies. This was to allow single frequency operation from different transmitters along airways.

This meant that totally new radios had to be developed. Unfortunately, the new radios are not totally compatible with the old ones used for ground stations at small aerodromes. The new airborne radios have to be used with both types of ground station, so the simplest method of doing that is to have two separate receivers inside the aircraft radio sets, and a means of switching between the two receivers.

Although the new ground stations are able to work with old aircraft sets when using 25 kHz spacing, the converse is not true. New aircraft sets have problems working with old ground stations. The pilot must switch his aircraft radio to 8 kHz spacing when operating above FL 245, but needs to return to 25 kHz when communicating with a ground station with an old radio, especially at short ranges. Unfortunately, changing the spacing usually involves going through a menu system rather than simply pressing a button.

Factors affecting VHF range

As we saw earlier, the formula for calculating the maximum range of a VHF signal is:

$$D = 1.25\sqrt{H_1} + 1.25\sqrt{H_2}$$

where D is in nautical miles, and H_1 and H_2 are in feet above mean sea level. Therefore, the factors affecting the range of a VHF transmission are as follows:

- Transmission power both at aircraft and ground station
- Height of the transmitter
- Height of the receiver
- Obstacles at or near the transmission site will block the signals or scatter them with inevitable attenuation
- Any obstruction in the line-of-sight between the aircraft and the ground station will have a similar effect to that above
- In certain circumstances the aircraft may receive both direct and reflected waves which may cause fading or even short-term loss of communication

Selective Calling System (SELCAL)

Pilots on long-haul flights used to have to listen to the radios all the time, waiting for their own callsign to alert them to a message for them. This was tiring, especially on HF frequencies with a lot of static as well as receiver noise. The SELCAL system allows pilots to mute the receiver until ATC transmits a group of two pulses. These pulses are designated 'RED x', where x is a letter corresponding to the audio frequency of the pulses transmitted as a modulation on the carrier frequency. Each code is allocated to a specific aircraft listening on the frequency. When the relevant code is received, it activates an alarm in the cockpit, either a light or a bell or both, telling the crew to de-select the mute function and use normal communications.

There are restrictions on the use of SELCAL. It can only be used if all the following conditions are fulfilled:

- The ground station is notified as capable of transmitting SELCAL codes
- The pilot informs the ground station that he intends to use SELCAL, and informs them of his codes
- The ground station does not raise any objection to the use of SELCAL
- A preflight functional check must be carried out satisfactorily
- If the serviceability is suspect, listening watch must be resumed

Internal communications (Intercom)

Most aircraft communications systems include an intercom facility. This basically consists of an amplifier which directly amplifies the input from each crew member's microphone. Intercom signals can therefore be received in every other crew member's headset, or a loudspeaker, at a similar strength to those amplified from external radio waves.

Because there are many external signal inputs coming into most cockpits, it is usual to combine the intercom system with all the other inputs in an audio control console. In this console, all the received signals from the radios and navigation aids may be selected for listening independently as required.

The volume controls on each individual control unit determine the actual volume of each signal in the pilot's headset. Figure 3.2 shows a modern audio control console for a light aircraft.

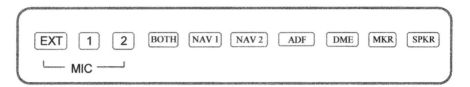

Figure 3.2 Audio control console.

Satellite communications (Satcom)

Although once a novelty, we now have satellite communications in many homes, giving us television pictures and sound. Aviation also uses satellites for communications, mainly via the International Maritime Organisation constellation INMARSAT. These satellites are positioned in 'geostationary' orbits very high over the equator, and provide communications by accepting transmissions of digital signals in the 6 GHz band. The signals from the satellites cover the whole of the earth between 80° North and 80° South, as can be seen in Figure 3.3.

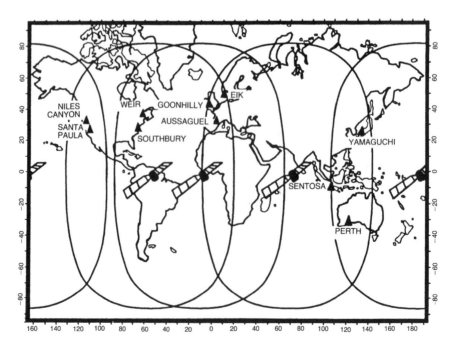

Figure 3.3 Satcom coverage and ground stations.

These signals are virtually unaffected by meteorological conditions or static. However, special aerials are required for transmission and reception on these frequencies. The satellites do not reflect the signals; they receive them and re-transmit them on different frequencies, so reducing the attenuation of the signal. Those re-transmitted to ground stations are sent in the 4 GHz band, those to aircraft in the 1.5 GHz band.

Ground stations are positioned in a network so that they service each of the four satellite regions or 'segments' and link into the conventional public and private telephone networks. This means that a pilot using the system is effectively using an ordinary telephone, as do his passengers from their seats! The aircraft satcom receivers operate on frequencies between 1544 and 1555 (ideally up to 1559) MHz. The aircraft satcom transmitters use frequencies between 1626.5 and 1660.5 MHz in ideal conditions, but generally between 1645.5 and 1656.5 MHz. Voice messages are digitised by the equipment using specific algorithms, which are laid down in ICAO Annex 10.

Search and rescue satellites

A further use of the satellite constellation is for search and rescue. All the INMARSAT satellites listen constantly for signals on the international emergency frequencies, and can alert SAR centres to emergency beacons carried by survivors. The earlier but still functioning international COSPAS-SARSAT system is dedicated to the provision of search and rescue facilities, and uses a different system of four polar orbiting satellites to cover all the globe. Cospas is the Russian name, Sarsat the US name for this joint venture.

One of these satellites can receive signals transmitted at 121.5 MHz, for example from a survivor's Personal Locator Beacon (PLB). The satellite re-transmits the signal to a ground station called a Local User Terminal or LUT, where the exact frequency received is measured and compared with the datum 121.5 MHz. The difference is the Doppler shift (see Chapter 13). That Doppler shift will only be the equivalent of the satellite's velocity if the transmitter is directly below the satellite's path. Any difference means the transmitter is to one side of the path. The maximum change comes as the satellite passes abeam the transmitter, and the variation of Doppler shift gives an indication of the lateral distance from the satellite's path, so a search area can be calculated.

Signals on 121.5 MHz can only be re-transmitted to LUTs which are in line-of-sight from the satellite. Signals from transmitters using 406.025 MHz, the international UHF search and rescue frequency, are sent as digital data streams, which include an individual identification signal. The data streams can be stored in the satellite for future transmission to a LUT, even though none is in line-of-sight when the original message is received. For this reason, 406 MHz emergency position indicating radio beacons (EPIRBs) are preferred for ocean voyages and flights.

ACARS

The Aircraft Communications Addressing and Reporting System (ACARS) is another system designed to reduce pilot workload in airliners. Much of the communication on airliner radios used to be on company frequencies, passing information about aircraft system serviceability, crew and passenger requirements, fuel state and requirements, and many other routine messages. As aircraft became larger and more complicated, these messages increased, usually requiring transmission during periods of high cockpit workload such as the descent into the destination.

With the advent of Flight Management Systems (FMS – see *Ground Studies for Pilots: Navigation*), most information which might need transmission already exists in digital form on the aircraft's computers. The ACARS can send that information from the FMS computer to computers on the ground. The crew can prepare their messages using the keyboard and scratchpad on the control and display unit (CDU) (see Chapter 16 and Figure 16.4 of this volume) if required, but many transmissions are automatic, requiring no extra workload on the flight crew. The ground computer can also send messages to the FMS for display on the scratchpad of the CDU. Information from other computers on the aircraft can also be sent, allowing ground engineers to monitor the aircraft systems while it is in flight, and arrange maintenance.

The ACARS can be compared to a facsimile (fax) machine. A data message can be delayed automatically until the frequency is vacant. It is compressed, so uses less time than a voice message. The ACARS equipment acknowledges messages automatically, and many aircraft have a printer to produce hard copy of the messages.

The ACARS uses a normal aircraft VHF radio set to send its signals, pulse-modulating the carrier to send digital signals. Usually such a set is dedicated to ACARS, but sometimes its use may be shared between the ACARS and normal communications by use of a VOICE/DATA switch. Frequencies 136.900 to 136.975 are reserved for datalink communications, but any frequency between 118.000 and 136.975 may be used at a frequency separation of 25 kHz. The frequency of 136.975 itself is reserved as a worldwide common signalling channel to announce the availability of VHF datalink services by a particular transmitter.

Legislation on the use of aircraft radio

The JAA regulations follow ICAO Standards and Recommended Practices from ICAO Annex 10, as made into UK law by the Air Navigation Order. The most important are listed below.

(1)   An aircraft transmitter may not be operated in the air or on the ground unless the radio is licensed and of a type approved by the regulatory Authority (the CAA in the UK).

(2)   An aircraft transmitter may not be operated in the air or on the ground unless the operator is appropriately licensed.

(3)   Radio transmitters in public transport aircraft must be capable of operation on any frequency notified for the route to be flown, and must be kept serviceable.

(4)   A continuous listening watch must be kept on the notified frequencies (which may be satisfied by SELCAL equipment).

(5)   Radio equipment must not cause interference to aeronautical communications or navigational services.

(6)   In a UK registered aircraft on a public transport flight, the pilot and flight engineer (if any) must not use a hand-held microphone whilst the aircraft is flying in controlled airspace below FL 150, or is taking off or landing.

Summary

HF   Allocated frequencies 2–22 MHz (only used up to 18 MHz)
Sky wave – Long range – Night frequency $\frac{1}{2}$ that of day – skip distance
AM voice or keyed
 	
SSB removes carrier AND one sideband from transmission, Receiver produces internal carrier to oscillate with received sideband
 	
SELCAL for voice communications
Receiver switched to standby – alarm activated by individually addressed code
Can only be used if: • ATC have published SELCAL facility
                                • Pilot informs ATC of his intention and codes
                                • ATC do not object to its use
                                • Preflight test satisfactory
 	
Digital HF   Datalink or voice – USB

VHF   Allocated frequencies 117.975–137 MHz
Direct wave – short range. Range $= 1.25\sqrt{\text{a/c height}} + 1.25\sqrt{\text{transmitter height}}$
SELCAL also available
Interference from reflected waves and tropospheric ducting
Frequency separation 8.33 kHz in modern equipment
International emergency frequency 121.500 MHz

ACARS   Datalink for company requirements
Manual preparation or automatic
Can use normal VHF frequencies or allocated
136.900–136.975 MHz

UHF   **SATCOM**   Geostationary satellites over equator act as relay sets
Quoted as capable to 80°N or S, four satellites cover the world to there
Different frequencies for transmission and reception at aircraft

EPIRB   406 MHz (really 406.025 MHz)
Individual coded digital messages
Can be stored for re-transmission later
Received by LUTs on ground

Sample questions

1. A single sideband transmission:

   a. Normally transmits the carrier wave also?
   b. Is used to reduce atmospheric noise received?
   c. Has an emission designator of A3J?
   d. Is usually in the SHF band?

2. At what maximum range would a VHF signal transmitted from an aircraft flying at 14 000 ft be received by another aircraft flying at 9000 ft:

   a. 190 km
   b. 212 km
   c. 267 km
   d. 494 km

3. If good HF communications are established at night, what should you do to maintain these communications the following day?

   a. Nothing
   b. Double the frequency

    c. Quadruple the frequency

    d. Halve the frequency

4. When may a crew use the SELCAL facility of an HF set?

    a. When ATC publish the facility is available

    b. When the crew specifically obtain ATC's permission to use it

    c. Both (a) and (b)

    d. Neither (a) nor (b)

5. What is the maximum latitude at which satellite communications are effective?

    a. 90°N or S

    b. 80°N or S

    c. 60°N or S

    d. 45°N or S

6. Which of the following is true about satellite communications for aircraft?

    a. Speech is digitised, several conversations can share one frequency

    b. Speech is digitised, each conversation has its own frequency

    c. Each satellite uses one frequency for all communications

    d. Satcom uses frequencies in the VHF band

7. What wavelength is most likely to be used for voice communications over 1000 miles during the day?

    a. 20 cm

    b. 2 m

    c. 20 m

    d. 200 m

8. What is the emission code of a single sideband voice transmission?

    a. P0N

    b. J3E

    c. A3E

    d. A9W

9. Which wave is used for VHF communications?

    a. Sky wave

    b. Surface wave

    c. Ground wave

    d. Space wave

10. Which emergency frequency is preferred for survival radio transmitters carried during flight over large ocean regions?

    a. 406.025 MHz
    b. 243.000 MHz
    c. 136.975 MHz
    d. 121.500 MHz

Chapter 4
VHF Direction Finding (VDF)

Introduction

Early in the history of radio, it became obvious that, if a station could determine the direction from which a signal was coming, that information could be very useful for navigation. Marconi had actually patented a method of radio direction finding in 1905. Of course, the military were the first to use it for their own purposes. Such direction finding was used by the UK during the First World War to monitor and locate German ships and aircraft, and in the Second World War to control fighter aircraft in the Battle of Britain, and to find U-boats in the Battle of the Atlantic. Even now, VHF direction finding provides a useful aid to navigation in the air and to identification for Air Traffic Control.

Principles

A standard half-wave dipole antenna receives signals from all directions equally, and has a horizontal polar diagram in the form of a circle, as in Figure 4.1(a). If two half-wave dipole antennas are held in a position half a wavelength apart, their reception and re-transmission characteristics are such that their horizontal polar diagrams form a figure 8 as in Figure 4.1(b). This means that if the pair of antennas are rotated, they will receive varying signal strengths from maximum to zero. Because there are two very sharp null points. it is easy to find the direction of an incoming signal. However, there are two possible directions, 180° apart, so a further pair of antennas can be employed as switchable reflectors (sense antennas) to affect the strength of signal received close to one of the nulls, as in Figure 4.1(c). An operator could note the direction of the null with the sense antennas switched off, turn back until he received the signal again, then switch in the sense antennas. The signal strength would either increase or decrease depending on the actual direction of the signal source. This set of four antennas was referred to as an 'Adcock' aerial.

This system is not efficient, requiring time and a dedicated operator. More modern systems use a number of fixed receiver directional elements arranged in a circle, pointing outwards. Each element will receive a slightly

(a)   (b)   (c)

Figure 4.1 DF aerial system.

different strength of signal from the aircraft. Electronic switching samples the signal strength of each of these elements in turn around the circle, very rapidly. This produces a sine curve of signal strength, one cycle per revolution. Figure 4.2 shows one of these systems.

At the same time, a generator in the system produces a reference signal at the same frequency as the sampling. The phases of the sampled signal and the reference signal are compared, and the phase difference equates to the direction from which the signal is coming, compared to the datum direction where the reference signal started. The display can be arranged to give either true or magnetic bearings, or both.

Figure 4.2 Doppler VHF Direction Finder (photo Fernau Electronics).

Older displays use a cathode-ray tube for indicating bearings. With this type of equipment, the direction of the received wave is displayed on the tube instantaneously as a trace and the bearing is read off against a scale. A short transmission gives enough time to read off the bearing, and if the aircraft is close to the station overhead the signal strength will decrease, as will the length of the 'spoke', improving the quality of a DF let-down service.

Some more modern units use liquid crystal digital displays, showing the required bearing in the form of numbers. This can be either in a separate display, or as a part of a larger display unit. Bearings can also be displayed as lines on the same display as a radar screen, and this is used by radar controllers on the ground to assist in identification of aircraft.

Services

A ground VHF direction finding station can give true or magnetic bearings when requested by a pilot. For brevity, the requests can be expressed either in plain language or in what we call the 'Q code', as follows;

- QTE = the aircraft's true bearing from the station
- QUJ = the aircraft's true track to the station (the opposite to QTE)
- QDR = the aircraft's magnetic bearing from the station
- QDM = the aircraft's magnetic track to the station (the opposite to QDR)

QTEs and QDRs are normally used in en-route navigation as position lines. QDMs are requested when a pilot wishes to home to the station. QUJ is generally only used as a stage in navigation calculations.

Bearings from three or more stations are used (in the UK normally on the emergency frequencies, which in the VHF band is 121.5 MHz) to give triangulation fixes. VDF stations are plotted on a central map. QTEs from each of them to a particular transmitting aircraft can be plotted on that map, and where the QTEs intercept is the position of the aircraft. The QTEs can be transmitted either verbally by telephone, or automatically by electronic means along telephone lines. The more QTEs that can be plotted, the more accurate the resulting fix. The fix can be transmitted to the pilot by an operator looking at that central map. Two bearings can give a poor quality fix, but three is regarded as the minimum acceptable for safety.

Direction finding stations do not guarantee their service; they can refuse to pass bearings to pilots if the conditions are poor or the bearings do not fall within the classification limits of the station. They will of course give the reasons for any refusal.

Classification of bearings

According to the judgement of the operator, bearings are classified according to their accuracy as follows:

- Class A – accurate to within $\pm 2°$
- Class B – accurate to within $\pm 5°$
- Class C – accurate to within $\pm 10°$
- Class D – less accurate than class C

The controller will pass the bearing with its classification in the form 'Your true bearing 247°, class alpha'.

Scope of the service

There are many automatic VDF stations intended purely to assist in radar identification for ATC purposes. These stations are not listed in the AIP for the obvious reason that they do not provide a normal DF service to aircraft. The stations that are listed in the AIP provide a normal 'homer' service. Generally the class of bearing is not better than class B. Automatic VDF stations should not be used as en-route navigation aids but their service is always available in an emergency situation or when other essential navigation aids have failed.

Factors affecting range

Being a VHF emission, the range will primarily depend on the height of the transmitter and receiver; the line-of-sight range. As we have seen, the maximum range may be found by the formula:

$$D = 1.25\sqrt{H_1} + 1.25\sqrt{H_2}$$

where D is the maximum range in nautical miles, H_1 is the altitude of the aircraft in feet and H_2 is the elevation of the station in feet, both above mean sea level. Under normal circumstances, using a station close to sea level, an approximation may be calculated quickly by a simpler formula:

$$D = 12\sqrt{F}$$

where D is again the maximum range in nautical miles, and F is the aircraft's flight level. This simpler formula is recommended for practical use.

Other factors, for example the power of the transmitter, intervening high ground, etc., will also affect the range. In addition, as explained earlier, the station may receive both the direct wave and a ground-reflected wave, in which case fading might be experienced or the signals may be lost completely for a time, until the aircraft changes its position.

The aircraft's attitude when transmitting may also affect the results. In general aviation, VHF communication transmission is vertically polarised. The best reception will occur when the signal arriving at the ground aerial is vertically polarised. If the aircraft attitude places the transmitter antenna in the horizontal plane, no signal will be received at the ground station. Normally, such an extreme will not occur, but range can be reduced as the polarisation alters from vertical.

Factors affecting accuracy

If an aircraft's transmission has been reflected by either uneven terrain or obstacles through its travel to the receiver, the aerial will receive the signals from a direction other than the original transmission. A similar effect occurs if the signals arrive at the aerial site directly, but suffer reflection from the ground or obstructions before entering the aerial itself. Therefore two factors affecting the accuracy of VDF are:

- propagation error
- site error – this is by far the most significant error in most systems

In addition, when the aircraft is nearly overhead the station, the signal will not be received at the station's aerial; where this occurs we call it the 'cone of no bearing'. A problem can also occur when two aircraft transmit at the same time. The VDF station receives signals from both at the same time, which changes the shape of the sine curve received, and the resultant calculated bearing, in proportion to the relative strengths of the two signals.

VHF let-down service

The VHF let-down service, available throughout the world, has the primary advantage that the aircraft does not require any specialist equipment to carry out a let-down (descent and approach to land). It is therefore useful if airborne navigation aids fail. The stations which provide this service are listed in the ENR (en-route) section of the AIP with their frequencies and callsigns. Details of the actual procedures are published in the AD (aerodrome) section of the AIP, and extracts appear also in various commercial publications, such as AERAD or Jeppesen charts.

Two types of procedure are in current use; the VDF procedure and the QGH procedure. Generally, the VDF procedure is available at all airfields with published VDF stations, but QGH is only available at airfields annotated as such in the AIP. Where both procedures are available, they will follow the same let-down pattern.

VDF let-down

For a VDF let-down, the pilot calls the station and requests 'VDF let-down'. The station gives the pilot a series of QDMs (also termed QDL) in reply to his frequent transmissions. The pilot uses these QDMs to orientate himself, making his own calculations, including allowance for wind, to achieve the published approach pattern for landing.

QGH let-down

With the QGH procedure, the pilot is given a series of headings to steer by the controller, who assumes responsibility for directing the aircraft in the pattern using QDMs from frequent transmissions. The aircraft is first homed in to the overhead at a safe altitude, then directed along an outbound leg. When safely 'established' on the outbound leg, the controller instructs the pilot to descend. After a calculated time, the aircraft is turned back 'inbound' towards the airfield.

On the inbound leg, the controller will give the pilot headings to steer to the airfield, and clearance to descend to minimum descent altitude, at which altitude the aircraft can continue to home to the overhead until the airfield is in sight. If the pilot reaches the overhead again without seeing the airfield, he must go-around and climb away for another attempt or divert.

Sample questions

1. On a VDF let-down, the controller passes a QDM of '127° class Bravo'. The 'class Bravo' means that the bearing is accurate to within:

   a. $\pm 1°$
   b. $\pm 2°$
   c. $\pm 5°$
   d. $\pm 10°$

2. VHF range for a VDF let-down is:

   a. $3 \times \sqrt{\text{height (ft)}}$
   b. $1.5 \times \sqrt{\text{height (ft)}}$
   c. $12 \times \sqrt{\text{height (ft)}}$
   d. Line-of-sight

3. QTE means:

   a. the aircraft's true track to the station
   b. the aircraft's true bearing from the station
   c. the aircraft's magnetic track to the station
   d. the aircraft's magnetic bearing from the station

4. When using VDF, errors:

   a. are mainly due to coastal refraction
   b. are nil, because the transmissions are line-of-sight
   c. occur mainly due to the siting of the receiver aerials
   d. occur mainly due to the signal propagation

5. The 'Q-code' for an aircraft's magnetic heading to steer to reach the station in zero wind is:

   a. QDM
   b. QDR
   c. QTE
   d. QUJ

6. If a pilot wishes Air Traffic Control to give him headings to steer during a let-down using VHF direction finding, what does he ask for?

   a. QNH
   b. QTE
   c. VDF
   d. QGH

7. If an aircraft is at 8000 ft, and the ground station is at sea level, what will be the maximum range at which the pilot can expect a QTE from the ground station?

   a. 80 nm
   b. 107 nm
   c. 144 nm
   d. 200 nm

Chapter 5
ADF/NDB

Introduction

It was appreciated quite early in the history of radio that direction finding would be a considerable aid to navigation, both sea and air. Commercial ground stations sent quite strong radio signals, and experiments in large aircraft with dedicated operators soon produced acceptable results. It was possible to receive signals from broadcasting beacons, which themselves were non-directional, and determine the direction from which the signals were coming.

The early system of airborne direction finding from Non-directional Beacons (NDBs) has remained basically unchanged, although the airborne equipment has progressed considerably. In fact, many NDBs have been recently introduced to assist with runway approaches at medium-sized airports, and there are moves to use them to transmit correction signals for the latest satellite navigation systems, as we shall see later.

Non-directional beacons

The successors to the old commercial stations are the aeronautical NDBs transmitting AM signals in the upper LF and lower MF bands, between 190 kHz and 1750 kHz. The wavelengths of the transmissions, between 1580 m and 170 m, make half-wave antennas rather large, so nearly all of the aerials used are much shorter, with components included to increase inductance and capacitance to electronically simulate longer antennas.

The advantages of these frequencies lie in the diffraction they suffer close to the earth's surface. Aircraft can receive surface waves if the direct waves are disrupted by obstructions or the curvature of the earth. In fact, coastal NDBs can be used by both ships and aircraft. There are of course disadvantages also, as readers can deduce. We shall look at these disadvantages later.

Beacons are situated along airways to guide controlled air traffic, on ocean coastlines to provide navigation assistance far out to sea, on airfields to provide homing, and under runway approach paths to lead aircraft safely and speedily on to the Instrument Landing System (ILS). Each of these positions requires different characteristics from their transmissions, mainly in the signal strength and consequent maximum reception range.

Emissions

We have seen that modulating a signal requires power, for an AM signal up to 50% above the basic carrier transmission power. For that reason, some NDBs, mainly those used for long-range navigation at ocean coasts, are unmodulated, with a short period of keying used to provide an identification message. These emissions are coded N0N A1A (the second part of the code refers to the keyed part). Unfortunately, to receive any part of the signal, the airborne equipment must employ a beat frequency oscillator (BFO) (described later in the chapter). In addition, during the breaks in the transmission required for the keying, there is no signal for direction finding. This type of emission is not recommended unless there is no alternative, but is common in France.

To reduce the N0N A1A disadvantages, many NDBs use a different modulation to provide the identification message. The carrier wave continues unmodulated for most of its duration, but when the identification is needed some of the power is used to amplitude modulate the carrier for identification. The resultant emission code is N0N A2A. The BFO is needed only for initial tuning.

NDBs which do not need the maximum signal strength can have their carrier wave amplitude modulated continuously. A BFO is not required to receive or identify this signal, whose emission code is A2A.

Rated coverage

NDBs are designed for particular purposes. The area within which their signals must be receivable is calculated, and within that 'rated coverage' the signal strength must be sufficient to give good reception and D/F. ICAO recommends a particular field strength and signal/noise ratio that should be receivable at the edges of that area. In the UK, the rated coverage is always equal to or greater than the 'protected range', which is described later and published in the En-route (ENR) section of the AIP.

Types of NDB

- **Locators** These are low-powered NDBs, usually installed as a supplement to ILS and located together with middle or outer markers. A locator has an average rated coverage of between 10 and 25 nm. Emission is usually N0N A2A, and they send identification signals every 10 seconds.
- **Homing and holding NDBs** These are intended primarily as approach

and holding aids in the vicinity of aerodromes, with rated coverage of about 50 nm. Their emission and 'ident' characteristics are similar to locators.

- **En-route and long-range NDBs** These provide en-route coverage along airways and a long-range bearing facility for ocean tracking and similar operations. The NDB at Cocos Island may be received at a range of several hundred miles. The recommended type of emission is N0N A2A unless the required rated coverage cannot be achieved, in which case N0N A1A emissions may be used. Identification is broadcast once every minute.
- **Marine NDBs** These are situated along coasts to provide navigation assistance for ships. They can be used by aircraft, but their reception and serviceability is not guaranteed by aeronautical publishers. Many of them have A2A emissions.

Simplified principles of airborne D/F

The original method of airborne D/F involved a 'loop' aerial to receive the signals. The shape of the 'loop' could be any one of several, including rectangular similar to Figure 5.1 below.

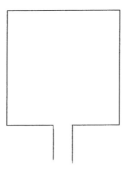

Figure 5.1 A loop aerial.

If a vertically polarised radio wave approaches the loop from one side, as in Figure 5.2, each arm of the loop will receive a slightly different strength of signal. This produces a slightly different electrical charge in each arm of the loop, and therefore an incentive for electrons to move, producing a small a.c. current.

Figure 5.2 shows a signal reaching the loop orientated in the same direction as the loop. If the loop is oriented across the signal, no signal will be received, and no current induced. The direction of this 'null' can be fairly clearly defined. At intermediate angles, the signal, and induced current, will be reduced in accordance with the cosine of the angle between the loop and

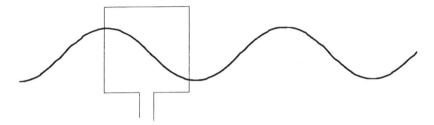

Figure 5.2   Signal at loop aerial.

the arriving signal. Unfortunately, because the induced current is a.c., there are two directions for the null, and four for each strength of induced current, as can be seen from the horizontal polar diagram on the left of Figure 5.3.

This can be resolved by using a second 'sense' dipole antenna to receive the transmitted carrier wave. Adding the amplified current from the loop to the induced current from the dipole produces a horizontal polar diagram as shown on the right of Figure 5.3(b). The shape of this goes under the name of a 'cardiod'. The null in this cardiod is less clearly defined, and occurs when the loop is orientated along the arriving carrier wave, because the signals from the two antennas are out of phase at that point, cancelling each other out.

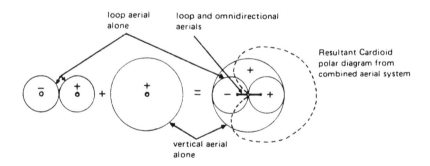

Figure 5.3   Loop aerial polar diagram.

ADF

Old airborne equipment required an operator to turn the loop and determine the original null, then switch in the sense antenna and turn it again to discover the correct direction. Progressively, equipment has improved to allow it to earn the correct expansion of the initials ADF – 'Automatic Direction Finder'.

The total signal from the cardiod can be amplified and fed to a motor which can turn the loop. The direction of turn depends on the phase of the total signal, which will automatically turn the loop towards the null position. When there is no total signal, the loop and its associated pointer indicate the direction of the incoming signal relative to the datum of the aircraft's longitudinal axis (fuselage).

A rotating loop is ungainly, and creates drag. The same effect can be obtained by using a fixed pair of loops arranged at 90° to each other. Each arm of the loops will receive a slightly different signal strength. The different strengths produce a magnetic field inside the loops, and a magnetic rotor inside or below the cage formed by the loops will align itself in this magnetic field. This is commonly called a 'goniometer'. The direction of the rotor can be reproduced by a signal selsyn in a remote indicator, to point towards the incoming signal.

Even more drag reduction can be achieved by using a flat, cross-shaped aerial with windings around the arms to give the same effect as loops. This aerial protrudes only a matter of centimetres above the aircraft skin.

The beat frequency oscillator (BFO)

In order for a pilot or operator to actually hear a signal, to either tune it or for identification, it must be at an audio frequency, between 300 and 3000 kHz. The NDB carrier waves are, of course, at a much higher frequency than that. A receiver can be fitted with a BFO, which the pilot can switch in to the receiver circuit when required.

The BFO is a device which produces a signal inside the receiver at a frequency of about 1000 Hz removed from the received wave. The received wave is compared with the produced signal, and the difference in frequency is converted to an actual audio signal at the same frequency which the pilot or operator can hear. The BFO must be switched on when manually tuning a N0N A1A or N0N A2A signal, and when identifying a N0N A1A signal. For direction finding, it should be switched off. In many equipments, the BFO is controlled by a 'tone' switch.

ADF controls

As shown in Figure 5.4, typical traditional ADF control boxes contain several switches. Having switched ON the equipment, an operator will first select the frequency of the desired NDB, and select the BFO with the tone switch if necessary. The ADF/ANT switch should be selected to ANT, allowing the operator to hear the signal from the sense antenna only for tuning and identification. If no signal is heard, check the volume control! After correct identification, the switch must be set to ADF for correct direction indication.

1: Frequency indicating dial.
2: Mode control switch.
3, 4, 5: Frequency selector
(100, 10, 1 and 0.5 kHz). 6: Tone
selector. 7: Audio GAIN control.
8: Transfer (TFR) switch – selects
one of the two chosen
frequencies and illuminates
appropriate indicator.

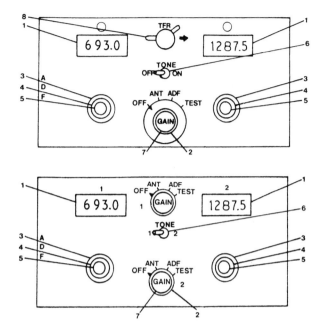

1: Frequency indicating dial.
2: Mode control switch.
3, 4, 5: Frequency selector
(100, 10, 1 and 0.5 kHz). 6: Tone
selector. 7: Audio GAIN control.

Figure 5.4 Typical ADF control units.

Presentation

RBI (Relative Bearing Indicator)

Old equipments use a fixed circular instrument face, marked in 360° from the nose of the aircraft. The pointer indicating the direction of the signal moves around the dial to indicate the direction relative to the aircraft's heading. The pilot or operator has to then make mathematical calculations to determine the magnetic or true bearing of the aircraft from the beacon in order to plot a position line. This instrument is called the 'Relative Bearing Indicator', or RBI, and looks similar to Figure 5.5. It is also sometimes called the 'Fixed Card Indicator', or 'Radio Compass'.

As you can see from Figure 5.5, when using the RBI, the pilot must add the RBI bearing to his own heading to find the true or magnetic bearing to the station. Assuming the upper aircraft is heading 150°, the reading on the RBI (the relative bearing) is 090°, so the bearing of the station from the aircraft is 090 + 150 = 240°. The bearing of the aircraft from the station is the reciprocal, i.e. 240 − 180 = 060°.

The lower aircraft is heading 090°. The actual bearing of the station from the aircraft to the station is 090 + 310 = 400 (but there are only 360 degrees in a circle so) subtract 360 = 040°. Again the bearing of the aircraft from the station is the reciprocal, i.e. 040 + 180 = 220°. Of course, whether the end

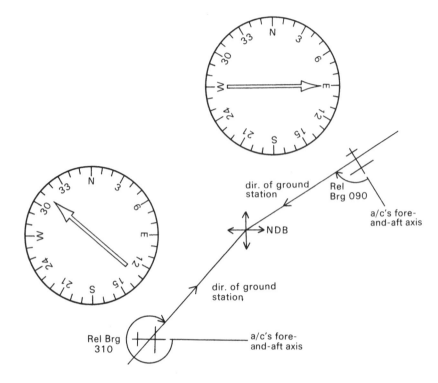

Figure 5.5 RBI.

bearing of these two calculations is true or magnetic depends on the aircraft's original heading. We can therefore use simple formulas to find bearings:

Magnetic heading plus RBI = QDM
True heading plus RBI = QUJ   (QUJ + 180° = QTE)

The RBI display was developed further. Most RBIs now allow the pilot to turn the indicator scale until the aircraft heading appears at the top. The pointer still shows the relative bearing of the arriving signal, but provided the pilot has correctly set the aircraft heading, it also indicates the actual bearing of the signal from the aircraft. The other end of the pointer (the tail of the needle) shows the bearing of the aircraft from the beacon.

RMI (Radio magnetic indicator)

The common display is now the Remote Magnetic Indicator or RMI, as seen in Figure 5.6. The scale is automatically orientated, like a remote indicating compass, to the earth's magnetic field. Again, the needle indicates the actual

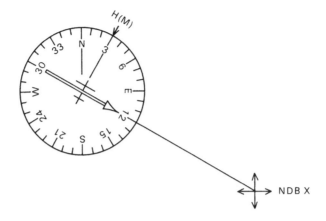

Figure 5.6 RMI.

bearing of the beacon from the aircraft, the tail indicates the bearing of the aircraft from the beacon, and the relative bearing of the beacon can be assessed from the position of the needle relative to the top of the instrument. Most RMIs have two needles, each of which can be selected to show information from an ADF or a VOR equipment.

An interesting and important fact should be noted about ADF indications on an RMI. The heading on the RMI may not be correct, for example the magnetic sensor may give false information. However, the relative bearing is fed directly from the ADF equipment, so will be correct no matter what the heading indication. We shall look at the use of the RMI in more detail later.

The information on the instrument in Figure 5.6 shows that the aircraft heading is 030°(M). The QDM is 120°, and the QDR is 300°. The relative bearing can be interpolated roughly, and confirmed by subtracting the heading from the QDM as 120 − 030 = 090°(R). The RMI often forms part of a horizontal situation indicator (HSI), and both are described later. The HSI can be mounted on its own in the panel, or can be electronically simulated on a modern display, either CRT or LCD. The information can also be used as an input for area navigation systems.

Use of the RBI

Some NDBs are still used to mark airway reporting points, and many more are used to guide aircraft in to runway approach aids. Pilots will normally be expected to fly to these NDBs, and to fly away from them on designated tracks.

Homing

To use the RBI to fly to a beacon, the pilot must tune and check the identification of the NDB. Once the ADF has locked on to the NDB, the relative bearing can be measured. The relative bearing can be added to the current heading to find the magnetic track towards the beacon (QDM), then he should turn his aircraft on to a heading which will take him towards the beacon. That heading must allow for the expected drift, so should be calculated before the turn, as quadrantal error (described later) will increase during the turn.

Once turned on to the calculated heading, the RBI will show the NDB slightly to one side of the nose. The relative bearing will equate to the drift; if port drift, bearing will be less than 360°, if starboard drift, bearing will be more than 000°.

If the bearing remains constant, the drift calculation was correct, and the aircraft will track to the beacon. If the bearing changes, however, as in Figure 5.7, alterations are needed. Drift was obviously wrong. Make a new estimate of drift. If the bearing moved further from the nose, too much was applied initially and you should turn towards the pointer. If the bearing moved

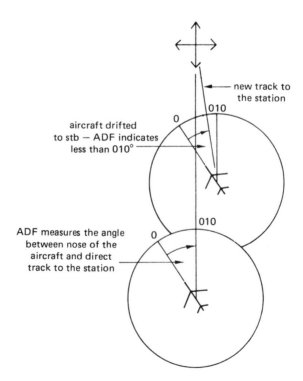

Figure 5.7   Homing.

towards the nose, initial drift was too little and you should turn away from the pointer.

Homing on a specific track

Complications can arise if you are trying to follow a specific track towards the NDB, as during a let-down. Initially, to arrive on the desired track, you should turn on to a heading more or less towards the beacon, but at an angle of 30° to the inbound track, as in Figure 5.8. The ADF needle indication should gradually 'drop' towards an angle of 30° (or 330° if the beacon is on your left). As it reaches that 30° 'cut', turn on to the calculated inbound heading (see Figure 5.8).

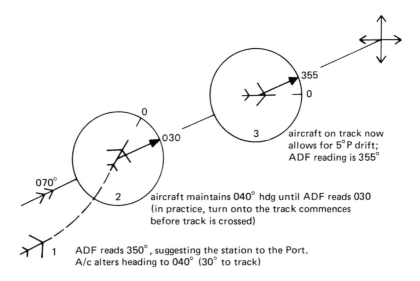

355
0

030

3   aircraft on track now
allows for 5°P drift;
ADF reading is 355°

0

070°

2   aircraft maintains 040° hdg until ADF reads 030
(in practice, turn onto the track commences
before track is crossed)

1   ADF reads 350°, suggesting the station to the Port.
A/c alters heading to 040° (30° to track)

Figure 5.8   Tracking to an NDB.

Once established on the inbound heading, if it is incorrect you can use a similar technique to return to the track, this time using a smaller angle, say 10°. However, before changing heading, calculate a new inbound one allowing for the changed drift.

Leaving a beacon

When leaving a NDB, things are much simpler. You can calculate a heading allowing for expected drift, and so you should. However, if you forget, or cannot estimate the drift, just fly along your intended magnetic track. After a short while, the needle will indicate the relative bearing of the beacon as being to one side of your tail. The amount it lies to the side is the amount of drift you are experiencing. Change your heading until the bearing is the same amount on the opposite side, and you have allowed for drift (see Figure 5.9).

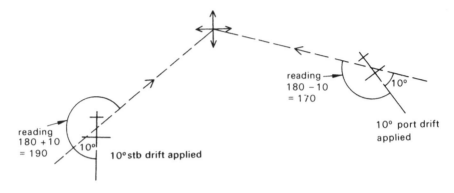

reading
180 − 10
= 170

10° port drift
applied

reading
180 + 10
= 190

10°

10° stb drift applied

Figure 5.9   Leaving the NDB.

Holding procedures

Holding procedures and let-downs using the RMI are considered later in this chapter. If the RBI is the only display available, more calculations will be required to follow the patterns, using the techniques above.

ADF and NDB errors

These are grouped together as system errors, although it is important, especially for examination purposes, to remember that the NDB is the ground station and the ADF is the aircraft equipment.

Quadrantal error

As mentioned earlier, the use of LF and MF frequencies does have dis-advantages. These produce errors in the system. However, this major error is not caused by the characteristics of the signal, but by the receiving aircraft itself. The waves arriving at the aircraft strike the whole aircraft. Parts of the aircraft reflect the waves, and many reflections arrive at the antenna, mixing with the direct waves. This affects the null signal, and the indication can suffer considerable errors. The greatest of those are commonly called 'quadrantal error'.

The error is at a minimum when the arriving signals are in line with the fuselage. It is also small if the signals come from the beam. However, a signal coming from 45° to the fuselage will produce the greatest quadrantal error, up to ± 20°. Individual ADF equipment can be 'swung' like compasses, but modern installations compensate for it when they are fitted by adjusting the current in the stator of the goniometer.

When an aircraft is banked, the wing area is exposed to the vertically polarised signal, and reflection is increased. Therefore, a turning aircraft can be expected to suffer more from quadrantal and other reflection errors, even though most of the error is compensated on installation. This is often referred to as 'dip' error.

Sky wave interference

Normally, the ADF will receive signals from the NDB as surface waves. The low frequencies are absorbed by the D layer of the ionosphere during the day. However, at night the D layer disappears, the other layers thin and rise, and the waves from the beacon can be refracted back to the receiver.

The receiving loop now has a second signal inducing current. Because this signal is received by the top and bottom of the loop, as in Figure 5.10, there is a current even though the loop is orientated at 90° to the incoming signal. The addition of the current induced by the sky wave can cause errors up to 30°.

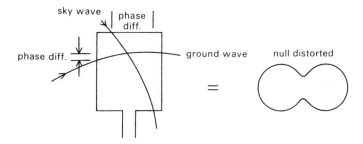

Figure 5.10 Sky wave interference.

Because the surface wave, by its propagation, is attenuated with range from the transmitter, the sky wave is a greater problem at long ranges from the beacon, where it is proportionately stronger than the surface wave. Partly because of this, but more because of interference from other beacons, all NDBs have protected ranges published in the En-route section of the AIP, within which the sky wave should have minimum effect during daylight. Protected ranges, however, are not valid at night.

Coastal refraction

The amount of diffraction and attenuation of the surface wave changes with the surface on the earth below. That implies that when a wave is over land it has a different speed (faster) to that when it is over the sea. At the coast, the speed change produces a bending of signal towards the land, where attenuation is greater. This bending does not happen when the signal is at 90° to the coast, but can have a considerable effect when the signal crosses at 30° to the coastline, as in Figure 5.11. As can be imagined, the effect is greatest when the wave crosses the coast close to the surface. High flying aircraft will suffer the effect less than low flying ones.

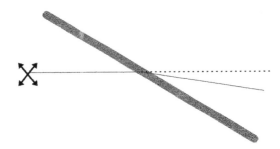

Figure 5.11   Coastal refraction.

If the NDB is on the coast, the refraction will happen immediately after the signal leaves the beacon, and an ADF will continue to give fairly accurate indications. Therefore, to minimise coastal refraction (sometimes called shore-line effect), aircraft should fly high and use NDBs as close to the coast as possible.

Static

Another problem with the chosen frequency bands for NDBs is that of static. Background static from electrical disturbances in the atmosphere can be received at very long ranges, and there are always thunderstorms taking place somewhere in the world, so as the NDB signal is attenuated, it becomes more difficult to distinguish from the static.

The relation between the signal and the noise from static is called the signal/noise ratio. ICAO recommends that the signal/noise ratio of NDBs should be a minimum of 3:1 within their rated coverage. Lightning from thunderstorms close to the aircraft can produce signals even stronger than the wave from the NDB. This can confuse the ADF, causing totally incorrect indications.

Power line interference

At the time of writing, electricity companies have proposed that data signals can be sent along their power transmission lines as frequency variations of the alternating current which currently passes along those wires. Should that become a common means of signal transmission, the power lines between the pylons would become effective antennas, transmitting the signals, or harmonics of them, into the atmosphere. These transmissions are likely to be at similar frequencies to those of NDBs.

Range

As we have seen earlier, the range of a surface wave depends on the frequency, the power transmitted, and the amount of power taken to modulate the carrier. The range of a sky wave in the MF and LF bands depends more on the state of the ionosphere than anything else, although power is very important also, because during the day the D layer will absorb most of the signal. Other transmissions at the same frequency can interfere with reception and D/F, so each NDB has a protected range published, within which interfering transmissions should be so much lower in signal strength than the intended signal that the receiver will ignore them. This protected range is not valid at night because interfering sky waves will be much stronger.

It is possible to receive a long-range NDB at up to 200 nm. However, during ocean crossings, pilots have been known to use coastal NDBs at longer ranges to give some idea of their position. Fortunately, modern navigation aids such as GPS allow us to dispense with the use of NDBs outside their protected range.

Factors affecting ADF range

- **Night effect** Sky wave interference reduces the maximum safe (protected) range to 70 nm
- **Transmission power** Range is proportional to the square root of the power
- **Frequency** Low frequency means less attenuation of surface wave so greater range
- **Emission** A2A has least range, N0N A1A greatest, because of the transmission power required for modulation
- **Terrain** Smooth terrain and especially a sea surface gives less attenuation so greater range

Accuracy

Most airborne equipment is capable of accuracies in the region of $\pm 2°$, but the combined accuracy of the system, including the NDBs, is $\pm 5°$ within the beacon's protected range.

Factors affecting ADF accuracy

- **Night effect** If outside 70 nm, the sky wave can cause errors of $\pm 30°$ at night
- **Terrain** In addition to coastal refraction, reflections from hills will change the signal direction. Flying high will reduce both problems
- **Static** Nearby thunderstorms not only cause static noise, but their electrical fields can be so strong they provide a stronger signal than the NDB transmissions.
- **Station interference** Do not use an NDB outside its protected range
- **Quadrantal error** This is calibrated in most aircraft, but will have an effect in turns (dip error)
- **Loop alignment** If the ADF system is not aligned with the longitudinal axis, errors will result

Warning

A major problem for pilots using the ADF is the lack of any warning device to tell him the system is unreliable or even failed. Even though it is important that pilots using the ADF constantly monitor the identification signal, that is no guarantee either that the carrier wave is strong enough for D/F, or indeed that it is present. There is also no indication that the ADF is functioning properly, although the NDB itself should be monitored by the state in which it is situated.

Summary

Frequency (Band)	190 to 1750 kHz (low MF and high LF)
Emissions	N0N A1A, N0N A2A, and A2A
Range	200 nm or more by day, do not use outside protected range by day, 70 nm at night
Range factors	Transmission power, night effect, frequency, emission, terrain
Accuracy	$\pm 5°$ within the beacon's protected range
Accuracy factors	Night effect, terrain, static, quadrantal error, station interference, loop alignment
Failure warning	None
BFO	Use for tuning and ident of N0N A1A, tuning only of N0N A2A, not for A2A

Sample questions

1. Which of the following are all factors affecting the accuracy of ADF indications?

    a. Night effect, tropospheric ducting, coastal refraction, hill effect
    b. Coastal refraction, static interference, loop alignment, quadrantal error
    c. Quadrantal effect, site error, night effect, station interference
    d. Thunderstorm interference, loop alignment, duct propagation, hill effect

2. Which of the following are all factors affecting the range of an ADF?

    a. Night effect, loop alignment, static interference
    b. Transmission power, static interference, type of emission
    c. Frequency, transmission power, night effect
    d. Terrain attenuation, frequency, loop alignment

3. Given all other factors equal, which of the following signals and emission types will give the greatest range from an NDB?

    a. A vertically polarised N0N A1A signal
    b. A horizontally polarised N0N A2A signal
    c. A vertically polarised A2A signal
    d. A vertically polarised N0N A2A signal

4. If an NDB emits a N0N A2A signal, what should the pilot do with the BFO switch?

    a. BFO on all the time
    b. BFO on during tuning and identification
    c. BFO on during tuning only
    d. BFO on during identification only

5. Which of the following frequencies might be used by an NDB?

    a. 40 kHz
    b. 400 kHz
    c. 40 MHz
    d. 400 MHz

6. An aircraft is heading 240°(M). The RBI indicates 030°. What is the magnetic bearing of the aircraft from the NDB?

    a. 030°
    b. 090°
    c. 210°
    d. 270°

7. A coastline runs North–South. At which aircraft position will coastal refraction most affect its ADF?

   a. 000° from the beacon
   b. 030° from the beacon
   c. 060° from the beacon
   d. 090° from the beacon

8. How can a pilot reduce the effect of coastal refraction when flying over the sea?

   a. By flying at high altitude and using NDBs well inland
   b. By flying at low altitude and using NDBs well inland
   c. By flying at high altitude and using NDBs close to the coast
   d. By flying at low altitude and using NDBs close to the coast

9. An aircraft is heading a constant 045°(M). On which magnetic bearing from the aircraft will the ADF be most subject to quadrantal error?

   a. 045°
   b. 075°
   c. 090°
   d. 135°

10. The protected range for an NDB is valid:

   a. at all times?
   b. only at night?
   c. only during the day?
   d. only over the sea?

11. Given all other factors equal, which of the following will be received at the greatest range during the day?

   a. A signal of frequency 250 kHz over the sea
   b. A signal of frequency 450 kHz over the sea
   c. A signal of frequency 250 kHz over the desert
   d. A signal of frequency 450 kHz over the desert

12. What is the maximum range at which ADF can be safely used at night?

   a. The published protected range
   b. Half the published protected range
   c. 70 nm
   d. 200 nm

Chapter 6
VOR (VHF Omnirange)

Introduction

An early directional transmitting navigation aid, radio range, operated in the MF/LF band. It served aviation for a period following the Second World War, and could be received on early ADF equipment. It had its limitations inherent with lower frequencies and at best it could produce only four fixed tracks to or from the beacon. A need for a more flexible and reliable aid soon became apparent, and VHF omnidirectional radio range (shortened to VHF omnirange or VOR) emerged as its successor. It was officially adopted by ICAO in 1960 as a standard short-range navigation aid.

VOR produces an infinite number of tracks to or from a beacon. It is practically free from static and does not suffer from night effect. Consequently, it can be used with confidence at any time throughout the 24 hours. Original indications were in terms of deviation to the left or right from a selected track. Nowadays, information may be fed to an RMI (as described in Chapter 7) to give indications of QDMs and QDRs. When used with co-located distance measuring equipment (DME), range and bearing information provides instantaneous fixes. VOR is also used as an input to modern computerised area navigation systems.

Principle of operation

The principle of VOR is bearing measurement by 'phase comparison'. In Chapter 5 we saw that an NDB transmits an omnidirectional signal and the aircraft's loop aerial measures the direction from which the signal is coming. A VOR transmitter sends two separate signals which can be compared to provide the directional information. No special receiver antenna is required in the aircraft.

Reference signal

The station transmits an omnidirectional horizontally polarised continuous wave signal on its allocated frequency between 108.0 and 117.95 MHz. The emission code is A9W. That signal is amplitude modulated by a 9960 Hz 'sub-carrier' which itself is frequency modulated at 30 Hz. This omnidirectional radiation has a circle as its horizontal polar diagram. At a given range from the transmitter, an aircraft's receiver will detect the same phase on all bearings around it.

As can be seen in Figure 6.1, the phase pattern produced is independent of the receiver's bearing from the station. In the receiver, the 30 Hz component of the transmission is used as a reference (or datum) for the phase comparison which will provide the bearing.

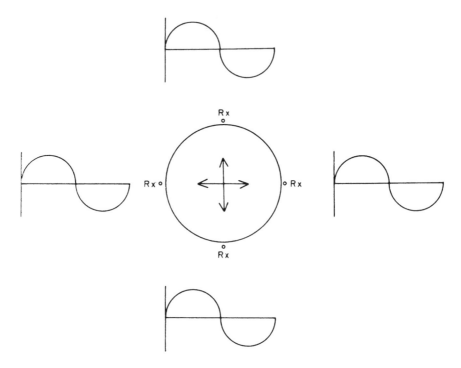

Figure 6.1   Reference signal – same phase in all directions.

Variable or directional signal

This is also transmitted on the station frequency, by an aerial which rotates either physically or electronically 30 times per second (30 Hz). The signal is less strong than the omnidirectional signal, so that together the signals appear to be a single carrier wave, amplitude modulated at 30 Hz to a depth of 30%. The horizontal polar diagram of this resultant signal is called a 'limaçon', and is represented in Figure 6.2.

An aircraft in a certain position will receive this rotating pattern as a signal with a 30 Hz amplitude modulation. The phase of this received AM signal will vary as the position of the receiver around the circle of rotation. This

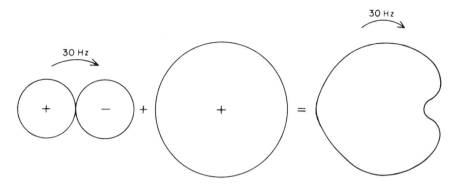

Figure 6.2   Derivation of limaçon.

phase can be compared with the phase of the FM omnidirectional signal, and displayed to the pilot. The transmitter is arranged so that the two signals are in phase when the receiver is in the direction of magnetic north from the transmitter. As the bearing from the transmitter changes, so does the phase of the variable signal. The phase difference from the reference signal equates to the magnetic bearing of the aircraft from the station. This is shown in Figure 6.3.

Station identification

The carrier wave has a keyed AM audio frequency signal to provide station identification at least once every 10 seconds, as recommended by ICAO. This identification consists of three morse letters transmitted at a rate of about seven words per minute.

  The carrier can also be modulated at audio frequencies by a voice signal. This can give mere station identification, or it can include more detailed information. Weather reports are frequently carried, but other AFIS (aerodrome flight information service) reports can also be transmitted.

Cone of confusion

ICAO recommends that the signals are transmitted at least up to 40° in elevation. In practice, modern equipment is capable of radiating signals up to 60° or even 80° above the horizon. That still leaves a gap overhead the station, in the form of a cone where no planned radiation takes place. However, while passing through this zone the receiver comes under the influence of weak 'overspill' (signals transmitted in incorrect directions, but usually very weak), causing confusion to the indications in the airborne equipment, which may change irregularly.

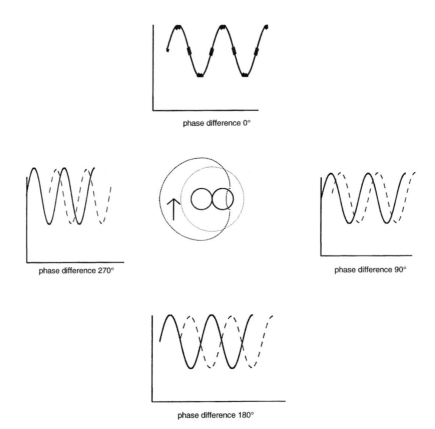

Figure 6.3   Bearing by phase comparison.

Ground equipment

Aerial system

The conventional VOR ground antenna can take several forms, but one of the most common is effectively a cylindrical cover enclosing a horizontal rotating dipole. Slots in the cylinder combined with the rotating dipole produce the limaçon-shaped rotating polar diagram. It is normally mounted above a room containing the power supply. A representation of such an antenna system is at Figure 6.4 below, and a picture of one surmounted by a DME antenna is shown in Chapter 14 at Figure 14.1.

Monitoring

Every VOR station has a monitor unit located in the radiation field, near the transmitter. This monitor receives the signal and compares it

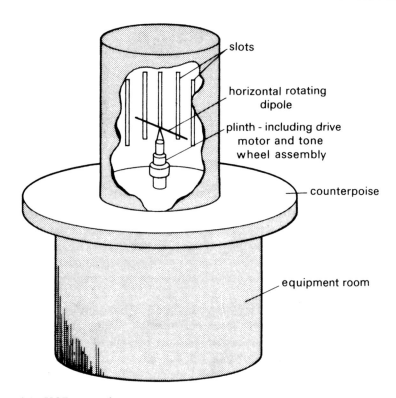

slots

horizontal rotating dipole

plinth - including drive motor and tone wheel assembly

counterpoise

equipment room

Figure 6.4   VOR ground antenna.

with known parameters. It will switch off the transmitter, or at least remove the identification and navigation signal, if any of the following take place:

- Bearing information is wrong by more than 1°
- Reduced signal strength (15% below normal)
- Failure of the monitor itself

Each station has a standby transmitter which will take over responsibility for transmission if the monitor switches off the main transmitter. This take over requires a certain amount of time, and during that time the signal can be received by airborne equipment. However, until the signal from the standby transmitter has stabilised and the monitor checks that it is correct, no identification signal is transmitted and the signal must not be used for navigation – considerable errors may be present. Especially when using a VOR station for airfield approach, pilots must listen to the identification code at all times in order to ensure safety.

Airborne equipment

The omnibearing selector (OBS)

The magnetic bearing from the station is called the 'radial'. There are an infinite number of radials, of course, but pilots normally refer to a radial as being of a particular whole number of degrees, so one can say there are 360 possible radials. Most VOR stations are positioned along airways, and an airway centreline is defined as a particular radial from one station until it meets the reciprocal radial from the next station along the airway. The changeover comes at the halfway point.

Because of this, the original display was the so-called 'omnibearing indicator' or OBI. The pilot has to select the desired radial manually, so the unit is often called the 'omnibearing selector' or OBS. The two expressions are frequently interchanged. To be pedantic, the OBS is only part of the unit, which includes a TO/FROM indicator and a LEFT/RIGHT Course Deviation Indicator (CDI) also.

The pilot selects the radial along which he wishes to fly, or its reciprocal, on the OBS as an intended track. The TO/FROM indicator, in the form of white triangles, tells him whether his selected track will take him closer to the station or further away (TO meaning closer). The vertical deviation (LEFT/RIGHT indicator) bar tells him whether the desired track is to his left or his right. Two examples of OBI are shown in Figure 6.5. The indicator on the right, the more common, has a track of 090° selected, whereas the indicator on the left has a track of 050° selected. The left-hand OBI has an extra horizontal bar, which is used for ILS signals as we shall see later.

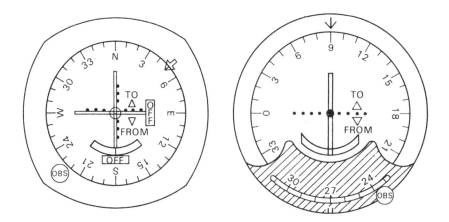

Figure 6.5 OBIs.

To/from indicator

It is sometimes difficult to interpret the OBI information. It must be remembered that everything depends on the selected track. As we have said, if that track will take the aircraft closer to the station, then TO will appear in the TO/FROM indicator. If the track will take the aircraft away from the station, the FROM triangle will appear. If the track will take the aircraft at an angle of between 80° and 100°, then there will be no indication of either TO or FROM. Figure 6.6 shows the TO/FROM indications that will appear when the selected track falls in the area shown. If the selected track is within ± 80° of the actual radial, TO will appear; if within ± 80° of the reciprocal, FROM will appear.

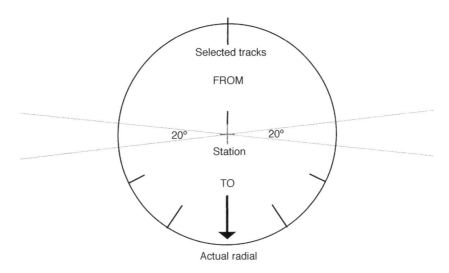

Figure 6.6   TO/FROM.

Course deviation indicator

The course deviation indicator (CDI) bar shows the pilot where his selected track lies, either to his left or his right. When the actual radial or its reciprocal lies outside 10° of the selected track, the bar will be positioned against a stop on the relevant side. As the aircraft moves closer to the selected track, the bar moves towards the centre until the aircraft is on the selected track. Dots on the indicator show how many degrees away from the selected track you are; as in Figure 6.5, some indicators have four dots on each side at 2.5° intervals, others have five dots at 2° intervals. Some indicators use a circle in the centre of the instrument to represent the first dot. In every case, the angular difference from the desired radial is always 10° when the needle is aligned with the furthest mark.

The important fact about the deviation indicator is that it will only tell the truth when the aircraft is trying to do what the TO/FROM indicator is indicating. In other words, if the pilot is trying to fly towards the station, he must set a desired track on the OBS which falls inside the **TO** area in Figure 6.6. If he is trying to follow a radial away from the station, he must select a track which falls inside the **FROM** area. If his selected track falls in the wrong area, the deviation indicator will indicate in the reverse sense.

Warning flag

Unlike ADF, the VOR equipment can detect many of the possible problems. A warning flag, usually red, will appear on the face of the instrument if any of the following is detected:

- no power or low power to the aircraft equipment
- failure of the aircraft's equipment
- failure of the ground station equipment
- failure of the indicator
- weak signals from either the reference or variable signal

The flag will also appear during tuning.

Characteristics

Frequencies

VOR shares the frequency band from 108.00 to 111.97 MHz with ILS signals. VOR signals in this band use all the frequencies with 'even' first decimals, for example 108.20, 108.25, 110.65. (ILS uses frequencies with 'odd' first decimals.) The frequency band 112.00 to 117.97 MHz is allocated only to VOR stations.

Range

Most VOR transmissions, being VHF, can be received at 'line-of-sight' ranges, calculated as in earlier chapters. However, because of the limited number of frequencies available, interference from other stations can be a problem. It has been known for sky waves from another station to interfere in certain circumstances. Listed in the AIP, each VOR station has a published 'Designated Operational Coverage', or DOC, within which the signal strength is guaranteed to be enough to avoid interference. This DOC is a cylinder of stated range from the station, up to a stated altitude, as shown in Figure 6.7. Do not use a VOR outside its DOC.

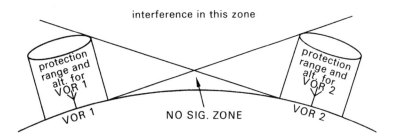

interference in this zone

protection range and alt. for VOR 1

VOR 1

NO SIG. ZONE

protection range and alt. for VOR 2

VOR 2

Figure 6.7 DOC.

Factors affecting range

(1) Transmission power.   The higher the transmitter power, the greater the range the signal can be received with adequate strength. En-route VORs with a power output of 200 W achieve ranges of around 200 nm. Terminal VORs (TVORs) normally transmit at 50 W, giving less range. The published DOC normally gives an indication of the transmission power available.

(2) Station elevation and aircraft altitude.   Because VOR transmissions are in the VHF frequency band, the theoretical maximum range depends on line-of-sight distance, as discussed earlier. (In practice, also as stated earlier, slightly better due to atmospheric refraction.) To calculate theoretical ranges for various altitudes, the VHF formula is used. This is repeated here for convenience:

$$D = 1.25\sqrt{H_1} + 1.25\sqrt{H_2}$$

where D is in nautical miles, and H_1 and H_2 are the altitudes of the station and the aircraft, respectively, in feet above mean sea level. However, as stated earlier, from a station close to sea level, the formula can be simplified for practical purposes to

$$D = 12\sqrt{F}$$

where F is the aircraft's flight level. As explained in *Flight Planning* in this series, each flight level is a 100 foot 'step' of pressure altitude (related to the standard atmosphere).

Accuracy

The transmitted signal is subject to errors, but for 95% of the time it must be at least better than $\pm 3°$. The errors due to the airborne equipment and interpretation are similar, but when all errors are combined, the accuracy of the indication will be within $\pm$ 5% for that 95% of the time.

Factors affecting accuracy

- **Beacon alignment**. The state operating the station is responsible for ensuring that the 000° radial is aligned with magnetic North. Of course, variation changes continually.
- **Site error**. Uneven terrain or physical obstacles in the vicinity of a VOR transmitter affect its directional propagation. Stringent requirements are laid down by ICAO regarding site contours and the presence of structures, trees, wire fences etc. Even the overgrown grasses affect the signals. This is sometimes called 'VOR course displacement error'. As mentioned earlier, the station monitor checks that this is kept to within $\pm$ 1°.
- **Propagation error**. After the signals leave the transmission site, they can still be reflected by the terrain over which they pass, and any obstructions in the path to the aircraft. These reflected waves further reduce the accuracy of the received signal.
- **Airborne equipment error**. Manufacturing inaccuracies produce small differences between the received signal and the actual displayed radial.
- **Pilotage error.** This is not a factor affecting the display, but when considering total accuracies of VOR signals, the difficulty in holding the aircraft on the desired radial is included in the calculations.

Using the equipment

The important fact about the OBI indications is that they are totally independent of the aircraft's heading. Only the OBS track is important.

Deriving a position line

To find a position line from the OBI, the pilot turns the OBS until the TO/FROM indicator shows FROM, and the deviation bar is in the centre. The resulting track shown in the OBS is the radial, or the magnetic bearing of the aircraft from the station. The reciprocal is the QDM to the station. Most equipment does not give instant readings – time and care are needed.

Homing

To home to the station, turn the OBS until the TO/FROM indicator shows TO, and the deviation bar is in the centre. The track shown in the OBS is the QDR. Turn on to that QDR, and allow for drift if you can. If your drift allowance is incorrect, the deviation bar will move to one side. Turn towards the bar by a sensible number of degrees, and hold that heading. Realign the CDI with your current track. If the bar moves again, turn towards it again through half your original change, and re-align it again. Continue to change heading by progressively smaller amounts and re-align the bar until the TO/FROM indicator either changes or fails to indicate.

Track following

When flying on airways, or making a VOR procedural let-down to an airfield, you will usually have to follow specific radials to or from a VOR station. To follow a selected track to or from the station, you must first arrive on that track. Select the intended track on the OBS, and look at the OBI. Confirm that the TO/FROM indicator shows what you expect! If the deviation bar is against the stops, select a heading which is towards the deviation bar but about 60° away from the desired track, as shown in Figure 6.8.

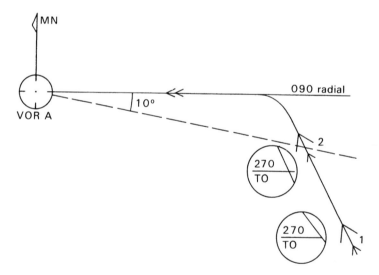

Figure 6.8 Finding the track.

Once established on a heading which will take you towards the desired track (270° in Figure 6.8), calculate a heading which takes account of the wind to track along the desired radial. At the same time watch the deviation bar. 10° before you reach the track, the bar will start to move. If it moves quickly, you are close to the station and should turn on to your calculated heading immediately. If the bar moves slowly, you are far away and should wait until the bar is within about 2° of the centre.

If you know your radius of turn and your range from the station, you can make a mathematical calculation to give a time to turn exactly on to the radial. Using the 'one-in-60' rule will tell you how many miles each degree of deviation equals. However, a little practice at watching the speed of bar movement and turning accordingly is equally as good and much less effort.

Once established on your calculated heading, maintain that, provided the bar is within half of its full deflection, and watch the bar. If it moves away from the centre, turn 5 to 10° towards it. Maintain that and wait. If the bar does not start to move towards the centre, double your original heading change. When it does return to the centre, turn back half your original change and hold that. Continue to 'bracket' the radial until you are satisfied you have allowed for the correct wind effect. As you close into the station, the bar will become increasingly sensitive. Do not attempt to follow the bar once you can actually see it moving.

Other displays

The VOR signals can be displayed in many forms, including hand-held combined communication/navigation radios. Modern airborne navigation computers can take the signals and use them to compare with other aids to find accurate positions without calculations from the pilot. Apart from this area navigation (R-Nav) facility, described in Chapter 16, many light aircraft displays use the remote magnetic indicator (RMI). A more modern version of a combination of the OBI and RMI, the horizontal situation indicator (HSI), is in use in larger general aviation aircraft and some airliners. Both RMI and HSI are described in Chapter 7.

Doppler VOR

VOR stations were needed to provide radials for all airways and most airfields. However, in many places the site errors could not be reduced to below the required ± 1°. For this reason, a different type of station had to be developed. This used the Doppler principle, as described in Chapter 13, but had to allow the use of standard airborne equipment without modification.

Aerial system

The principle is similar to that described earlier in Chapter 4 on VHF direction finding. A central aerial transmits an omnidirectional carrier wave, amplitude modulated at 30 Hz. In its simplest form, a series of antennas arranged in a circle around the central aerial transmit a second signal. This second signal consists of a separate continuous wave, 9960 Hz displaced from the station frequency, transmitted in turn from each of 52 antennas on the circumference 30 times per second. This makes the signal appear to rotate at 30 Hz. The size of the aerial array produces the same effect as that of the FM signal from a basic VOR.

Because the central aerial transmits the AM signal, and the rotating signal is FM, the rotation is anticlockwise. This means that the airborne equipment receives the same signals and phase difference as from a normal VOR station. A DVOR aerial is shown at Figure 6.9.

Figure 6.9   DVOR ground aerial.

Test VORs

These are installed at certain aerodromes to enable pilots to test the airborne equipment during preflight checks. The transmitters are called VOTs, and their frequencies are published in the AIP. To test the airborne equipment from any position on the aerodrome, tune in to the frequency and centralise the deviation bar. OBS counters should indicate 000° FROM or 180° TO. If they do not indicate correctly within ± 4°, the equipment requires servicing. Normally, a pilot would test his VOR equipment by selecting the nearest station, checking the identification signal, and comparing the radial indicated with his known position.

Summary

Frequency (Band)	(VHF) 108.00–111.95 MHz using even decimals, 112.00–117.95 MHz using all.
Emissions	A9W
Range	VHF formula – $12\sqrt{F}$ (flight level), or accurately $1.25\sqrt{H_1} + 1.25\sqrt{H_2}$ nb DOC
Range factors	Transmission power, station elevation, aircraft altitude
Accuracy	$\pm 5°$ on 95% of occasions
Accuracy factors	Beacon alignment, site error, propagation error, airborne equipment error, pilotage
Failure warning	Warning flag appears if: • Low signal strength • Airborne equipment failure • Ground equipment failure • Indicator failure • Low or no power • Tuning in progress
Test VOR	VOT – preflight check, 000° FROM or 180° TO, $\pm 4°$

Sample questions

1. An aircraft's VOR is tuned to a station. When 025° is selected on the OBS the deviation bar becomes central and the TO/FROM indicator indicates TO. What radial is the aircraft on?

   a. 025°(T)
   b. 025°(M)
   c. 205°(T)
   d. 205°(M)

2. In question 1, if the ADF is tuned to an NDB on the same site as the VOR, and the RBI indicates a constant 011° on ADF, what drift is the aircraft experiencing?

   a. 11° port
   b. 11° starboard
   c. 14° port
   d. 14° starboard

3. In question 2, what is the aircraft's heading?

    a. 011°
    b. 014°
    c. 036°
    d. 039°

4. If a VOR station in the AIP has 100/30 000 written beside it, what does that mean?

    a. An aircraft can receive its signal at 100 km if it flies at 30 000 ft
    b. An aircraft can receive its signal at 100 nm if it flies at 30 000 ft
    c. Aircraft should not use the beacon on the 100° radial below 30 000 ft
    d. Aircraft should not use the beacon outside 100 nm or above 30 000 ft

5. To minimise quadrantal error from a VOR station, a pilot should:

    a. Choose a VOR station on the aircraft's beam?
    b. Choose a VOR station on the aircraft's nose or tail?
    c. Both (a) and (b)?
    d. Neither (a) nor (b)?

6. Doppler VOR compares a reference signal with a variable signal by:

    a. measuring the phase difference between them
    b. measuring the time difference between receiving them
    c. measuring the amplitude difference between them
    d. measuring the doppler shift between them

7. Which of the following are all factors affecting VOR range?

    a. Transmission power, aircraft altitude, night effect
    b. Station elevation, site error, static
    c. Station elevation, aircraft altitude, transmission power
    d. Night effect, station elevation, coastal effect

8. Which of the following are all factors affecting VOR accuracy?

    a. Beacon alignment, site error, propagation error
    b. Airborne equipment error, propagation error, coastal effect
    c. Pilotage, night effect, airborne equipment error
    d. Site error, night effect, beacon alignment

9. What is the emission pattern of a VOR signal?

   a. A3E
   b. A8W
   c. A9W
   d. J3E

10. Which of the following is the expected accuracy of a VOR signal?

   a. $\pm 5°$, all the time
   b. $\pm 3°$, 95% of the time
   c. $\pm 3°$, all the time
   d. $\pm 5°$, 95% of the time

11. Which of the following indications would be acceptable if you tuned the VOR to a station marked 'VOT' during the pre-takeoff checks?

   a. 180°, FROM
   b. 004°, FROM
   c. 359°, TO
   d. 175°, TO

12. The modulation of the fixed signal of a Doppler VOR station is:

   a. Frequency modulated at 30 Hz?
   b. Frequency modulated at 30 MHz?
   c. Amplitude modulated at 30 Hz?
   d. Amplitude modulated at 30 MHz?

13. Which of the following would not trigger the warning flag on a VOR indicator?

   a. Site error becoming greater than 1°
   b. Standby station on line
   c. Ground monitor failure
   d. Outside usable range from the station

14. From the OBI indications given, what is the correct interpretation if the aircraft is heading 270°(M)?

   a. Aircraft is 3°N of the 280 radial, heading towards the station
   b. Aircraft is 3°S of the 280 radial, heading away from the station
   c. Aircraft is 3°N of the 100 radial, heading towards the station
   d. Aircraft is 3°S of the 100 radial, heading away from the station

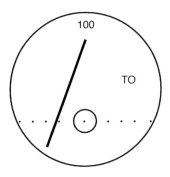

Chapter 7
Radio Magnetic Indicator (RMI) and Other VOR Indicators

Introduction

The Radio Magnetic Indicator or RMI has been mentioned earlier. As its name suggests, it is only an indicator and not an independent navigation aid. It accepts relative bearings from the ADF receiver and phase differences from the VOR receiver, and displays both on the face of a direction indicator aligned with the aircraft's heading. From these, QDMs and QDRs can be read.

It employs a rotating scale card calibrated in degrees and aligned with the aircraft's remote indicating compass. Thus, on the indicator at the 12 o'clock position, the aircraft's magnetic heading can be read off against a heading index. QDMs are indicated by two concentric pointers of different shape, each of which may be energised simultaneously by two like or unlike aids.

Figure 7.1 shows a typical RMI. The thin pointer is coloured red and the wide pointer is coloured green. By convention, the red pointer is called the number one needle and the green pointer number two. Again by convention, number one needle is usually used for ADF, and number two for VOR

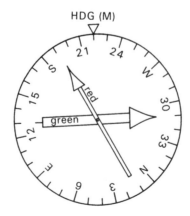

Figure 7.1   Radio magnetic indicator.

Figure 7.2   Electronic RMI (courtesy of Meggitt Avionics).

indications, but as shown in Figure 7.2 below, that is by no means always the case. In the extremely unlikely event that the main compass or its drive to the RMI fails, only relative bearings can be deduced from the RMI. Figure 7.2 shows a modern representation of a RMI on a 'glass screen' secondary navigation display. The same information is available, with the addition of other data such as DME ranges (see Chapter 14).

Warnings

Flags on the face of the instrument indicate the lack of a safe usable signal, as described in Chapter 6, from any VOR controller selected for indication on the RMI.

ADF relative bearings on the RMI

In the illustration in Figure 7.3, an aircraft on heading 030°(M) has tuned to NDB 'X'. The ADF gives a relative bearing of 090°(R). The measuring datum on the indicator is 000, that is the fore and aft (longitudinal) axis of the aircraft. From Figure 7.3 we can see that the QDM to the NDB is 120° and thus the relationship between a relative bearing and a QDM is the aircraft's heading. That is, as we have seen earlier,

Aircraft heading	(090)
+ relative bearing	(030)
= QDM	(120)

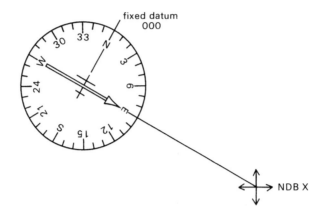

Figure 7.3   Relative bearing 090°.

In an RMI, this addition of the heading to the relative bearing is auto-matically carried out by adjusting the measuring datum. By measuring the pointer's indications from the aircraft's magnetic heading as its datum, as in Figure 5.6, the aircraft's QDM can be read off the numbers on the scale. Notice that the RMI displays a relative bearing with reference to the aircraft's magnetic heading as displayed on the heading index. If the pilot wishes to convert his displayed QDM to a true bearing from the NDB (QTE) in order to plot a position line, he must apply the magnetic variation which applies at the aircraft, not at the beacon.

VOR phase display on the RMI

As we have seen, VOR airborne equipment receives the phases of two different signals (the reference signal and the variable signal) and derives the QDR by measuring the difference between the two. The reciprocal of the QDR is the QDM. The RMI itself actually displays relative bearings. Therefore, the VOR phase difference is converted to give a relative bearing before it is displayed. This is achieved by use of a differential synchro in the VOR navigation unit which subtracts the aircraft's magnetic heading, as measured by the compass sensor unit, from the QDM before the bearing is fed to the RMI. The display itself then adds the aircraft heading again, so that the pilot can read it.

Magnetic variation

In Figure 7.4, the aircraft is heading 030°(M) and the RMI indicates 120° QDM from VOR 'Y'. Even if the magnetic variation at the VOR station and the aircraft are different, the indicator will continue to display the

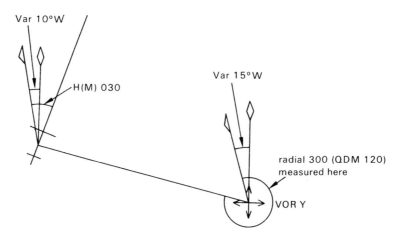

Figure 7.4 Variations at VOR station and aircraft.

actual magnetic radial from the station (QDM), which can be plotted directly.

We can confirm this as follows:

QDM received from VOR	120	(using variation at the station)
– aircraft heading (M)	030	(using variation at the aircraft)
= relative bearing	090	
+ aircraft heading (M)	030	(using variation at the aircraft)
= indicated QDM	120	(using variation at the station)

Thus, first the magnetic heading is subtracted, and then the same value is added. These two operations mutually cancel each other out and the resultant indication is affected only by the variation at the VOR station. Therefore, when converting VOR QDMs into QTEs for plotting, variation at the VOR station should be applied.

This situation has led to various questions in examination papers in the past. There is little doubt that similar questions will be asked in the future, so we shall investigate some possible scenarios.

Discrepancies in indications

1. If the magnetic variation at the aircraft is different from the variation at the VOR station, the indication of relative bearing is incorrect, but the indication of QDM is correct.

In Figure 7.5, aircraft A and B are both on a magnetic heading of 030°. The variation at A's position is 20°W, the variation at B's position, and also at the

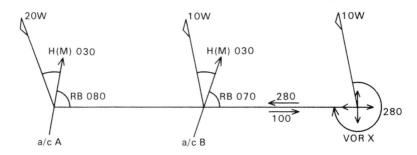

Figure 7.5   Variation effect.

station, is 10°W. Both aircraft are on a QDM of 100° from the VOR at X. From this figure, it will be apparent that the relative bearing of the station from aircraft A is 080°(R), and from aircraft B is 070°(R).

Both aircraft calculate their QDM as follows:

Aircraft A:	QDM	100	
	– Hdg (M)	030	
	Rel brg	070	this will be read off against a fixed RB scale, its true relative bearing is 080°
	+ Hdg (M)	030	
	= QDM	100	
Aircraft B:	QDM	100	
	– Hdg (M)	030	
	Rel brg	070	indicates the correct relative bearing (same variation)
	+ Hdg (M)	030	
	= QDM	100	

2. If an NDB is located on a VOR site, QDMs from the VOR and the NDB as displayed on the RMI will be different if variation at the aircraft is different from the variation at the station.

In Figure 7.5, we have already seen that the relative bearing indicated from the VOR in aircraft A would be 080°, whereas the actual relative bearing should be 070°. That relative bearing of 070° will be indicated correctly by the ADF needle, so in this case there will be a discrepancy between the two needles.

3. Convergency between the aircraft and the station will produce errors in relative bearings equal to the value of that convergency, although QDMs will be correct.

Because VOR range is relatively short, and relative bearings are little used at long range, this fact is of only academic interest, so it is not proposed to prove it here.

Plotting position lines (QTEs) from QDMs is covered in *Ground Studies for Pilots: Flight Planning* but can be summarised as follows:

QDM derived from NDB – apply aircraft variation and convergency and plot the reciprocal

QDM derived from VOR – apply station variation and plot the reciprocal

Advantages of RMI

- QDM/QDR are indicated continuously and read off directly. The tail end of the pointers indicate QDRs
- Using two stations, instantaneous fixes can be obtained
- RMI indications provide a useful guide when initially joining a radial for VOR homing
- The indicator itself can be used for homing
- Magnetic headings can be read off together with QDMs
- Approximate relative bearings may be assessed visually, or read off against an outer RBI scale

Use of the RMI

Homing to a VOR station

Having tuned the station, identified it, and ensured its signal is selected to one of the RMI pointers, the selected needle will point to the station, indicating the QDM. The pilot should calculate the drift he will expect on that track, for example 5° left, and turn his aircraft until the needle is indicating that number of degrees to the same side of the heading pointer, in this example 5° to the left, and maintain that heading. If the needle moves during the homing, the pilot must adjust the heading as in the case of ADF homing in Chapter 5.

Homing to an NDB

The same technique should be used, but because ADF dip will affect the needle during a turn, the pilot must decide his intended heading before he turns towards the NDB.

Tracking to a VOR station

The OBI is designed for this purpose. However, many pilots may have to join and follow radials using only a RMI. The easiest technique to join a radial and follow it to a station (for example joining and following an airway) is as follows.

To join a track to a station, first tune and identify the station, and select it on one of the pointers. Look at the needle which is pointing towards the station. Use a pencil or similar object (finger?) to show the desired track. Imagine the pointer turning the *shortest* way to line up with the pencil. Turn the aircraft the OPPOSITE way until you are heading 90° to the pencil. Confirm that the needle is actually moving closer to the pencil. Before you reach the desired track, you will have to anticipate and turn on to the desired track while the needle has some degrees to go before you reach it. Knowing range from the station and your radius of turn you can calculate the point to start your turn (from the '1 in 60' rule), but with practice pilots assess how quickly the needle is moving, and turn with a 'lead angle' depending on that rate of movement.

Look at Figure 7.1. Assume the pilot wishes to follow the 080° radial inbound. With the station selected on No 2 (green) needle, he should turn on to 350° to intercept that radial. Of course, there is often no need to join the desired radial at the closest point, in which case any heading which brings the aircraft on to the radial before the station may be acceptable. However, Air Traffic Service Units prefer aircraft to join airways at right angles.

The heading required to maintain the desired track obviously must be adjusted to take account for drift. This must be calculated before you turn on to it. In the above example, if you expect to have 10° of right drift, you should turn on to a heading of 250° before you reach the desired radial of 080°. When steady on the intended heading, the needle will point in the same direction as your expected drift. If that angle remains the same, the aircraft will track to the station. If, however, the RMI needle moves while you are approaching the station, you need to adjust your heading. Remember you are trying to fly on a nominated radial (tail of the needle) inbound. Alter heading by about 5° from the actual tail position to the desired tail position. When back on the radial, turn back by about 2°, and continue to monitor your radial. Obviously, if 5° does not work, you must make a larger alteration, but remember to alter from the present RADIAL (tail position) to the desired radial.

Tracking from a VOR Station

When overhead a VOR station it is usual to track outbound on a specified radial, such as an airway centreline. Before reaching the station, calculate the drift and heading required on the outbound track, and when overhead (when the needle 'falls' past a radial 90° to the inbound one) turn on to the calculated heading. When the instrument settles down, the heading at the top of the RMI should be offset into wind from the displayed radial by the amount of calculated drift.

If the needle settles in a different position, you need to adjust your position to reach the intended radial, then follow it. Again look at the tail of the needle, which is now near the top of the instrument. Adjust heading about 5° from the actual radial towards the intended radial. When the radial is correct, turn back to the planned heading, unless you have already proved that that is wrong by a previous correction, in which case turn back about 2°. Again, if the small correction of 5° is insufficient, make a larger alteration.

As in the case of following an ADF RBI needle from a beacon, if you have been unable to assess drift beforehand, it is possible to find approximate drift using the RMI. Set off from overhead the station on a magnetic heading equating to your planned radial itself. After a short while, the aircraft will have drifted off track, and the radial will have moved, by a certain amount corresponding to the drift experienced. Add that drift to the radial, turn on to that, then adjust heading to regain the correct radial as above. However, beware; airways are only 10 miles wide!

It is possible that you may have to join a nominated radial and track outbound. Use the same original technique as described above to join a radial inbound, but obviously turn away from the station when you approach the nominated radial.

Procedure turns

In many approach procedures, a pilot is required to change direction by 180°, first flying outbound on a radial, then turning round and flying back towards the station on the same radial. ICAO Document 8168 Vol. 1, *Procedures for Air Navigation Services – Operations* or 'PANS – OPS', describes the procedure to be adopted.

Ideally, the procedure turn involves starting from an established track on the radial, then turning away through an angle of 45°. Maintain that track for 75 seconds (1 minute for light aircraft), then make a continuous turn back on to the reciprocal of that last track for a short time until a final turn can be made on to the radial in the opposite direction to the original track, using the techniques described above. Figure 7.6 represents the standard procedure

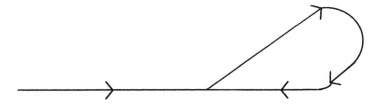

Figure 7.6   Procedure turn.

turn. An alternative method is to turn away from the original track by 80°, then immediately turn back in the opposite direction through 260° to line up with the reciprocal of the original track. As in all procedures, the bank angle used is to be 25°, or whatever gives a rate 1 turn (180° per minute), if less.

The holding pattern

Before or during an approach procedure, Air Traffic Control will often require a pilot to delay. This is usually done by asking him to 'hold' at a point. ICAO Document 8168 also lays down a procedure for this, described as a 'holding pattern'.

The pattern consists of two straight parallel tracks joined by two turns, as shown in Figure 7.7. One track, towards the point which is usually defined by a navigation aid such as a VOR station or an NDB, is called the inbound track, and is in turn usually defined by a VOR radial or a desired track to the NDB. The track parallel to the inbound track is called the outbound track, and is 180° different. Most holding patterns involve all right turns (a right-hand pattern), but a left-hand pattern, if required, is the mirror image.

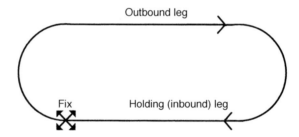

Figure 7.7 Holding pattern.

The purpose of the hold is for the aircraft to stay in the pattern until ATC allow it to start or continue its approach. Ideally, 4 min (6 min above 14 000 ft) should elapse between consecutive passes over the holding point, and the aircraft should follow the 'holding leg' or inbound track. Therefore, if there is any wind, timing and heading should be adjusted during the holding pattern to try to achieve these ideals. However, the twin aims of the pattern are to keep the holding aircraft within a certain area while it is holding, and to approach the fix along the holding track, so the procedure described below emphasises these aims rather than the ideals.

ICAO Document 8168 lays down that the maximum indicated airspeed during the pattern is 230 kts at or below 14 000 ft, 240 kts up to 20 000 ft, 265 kts up to 34 000 ft, and Mach 0.83 above that. In turbulent conditions, and with the permission of Air Traffic Control, speed may be increased to a maximum of 280 kts or Mach 0.8, whichever is greater, at any altitude up to

34 000 ft. Turns are again to be at rate 1 (180° per minute), or a maximum of 25° of bank.

Any wind component parallel to the holding leg will affect the timing. In order to achieve as close as possible to the ideal timing, a pilot should adjust his outbound timing, which in still air would be one minute. As required by the procedure, he starts timing the outbound leg as he passes abeam the fix, or when his wings become level after the turn overhead the fix if later. If there is a headwind on the holding leg, he must shorten his outbound leg by one second per knot of wind. This will compensate for the wind effect on the whole pattern. If the holding (inbound) leg has a tailwind, he should extend the outbound leg, again by 1 second per knot.

A crosswind component will affect the shape of the pattern. The pilot must calculate the drift expected along the holding leg to decide his inbound heading. He should then apply three times that drift (in the opposite direction) on the outbound heading to take account of the drift during the turns. Looking at Figure 7.7, if TAS is 180 kts, the inbound leg is 270°, and the wind is 315°/30 kts, on the holding leg there will be a headwind component of approximately $30 \times 0.7 = 21$ kts, and an expected drift of 10° to the left. (See *Ground Studies for Pilots: Flight Planning* for a reminder of the techniques used to calculate these.) The pilot should therefore steer 280° on the inbound track, and 060° on the outbound. He should maintain his outbound track for $60 - 21 = 39$ seconds from the time his wings become level after the overhead turn.

To fly a holding pattern on the RMI, some mental agility is required. The holding leg is no major problem, as the desired track and drift can be seen. The DI or compass heading bug, if available, should be set on the intended outbound heading before turning on to that leg. The RMI needle will indicate when the aircraft is abeam the fix for timing to start. However, the needle will have little relevance to the pattern after that until about halfway through the turn on to the inbound track. (Some pilots, however, claim that they can assess the progress of the pattern from the RMI needle indication at the end of the outbound leg, and it is possible by using the 1 in 60 rule.)

If the fix is a VOR station, the needle will give an indication, as the turn nears completion, of whether the pilot must roll out of the turn early, or continue it beyond his intended heading. If the fix is an NDB, dip will affect the reading, and experience and practice are required to make adjustments to the roll-out heading. The RBI requires a similar technique, but more mental agility to make good the inbound track, as explained in Chapter 5 on NDB.

An OBS makes the inbound track to a VOR easier. The course pointer of a HSI (described later in this chapter) is useful for the inbound track, its heading bug for the outbound heading, and the radial pointer shows when timing should start.

Joining the holding pattern

Pilots do not always approach the holding fix along the holding leg. In other cases, PANS-OPS lays down the correct way to join the pattern. If the aircraft approaches the fix in the same general direction as the holding leg, it is safe and sensible for it to turn on to the outbound track as soon as it arrives at the fix. This is called a 'sector 3' join, or sometimes a 'direct' join. In Figure 7.8, the boundary of sector 3 is shown. As can be seen, the boundary line is at an angle of 70° to the holding leg, although ICAO allows flexibility of up to 5° in the sector boundaries.

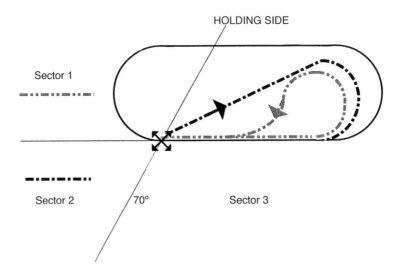

Figure 7.8 Joining the holding pattern.

If outside the boundary of sector 3, the aircraft is flown over the fix, and then an arrival pattern is flown to return to the fix and start the hold proper. These arrival patterns must be flown inside the holding pattern, so there are two options, depending on the direction of initial arrival.

If the arrival track is in the area of Figure 7.8 marked 'sector 2' it requires less than 70° of turn on to the outbound leg and comes from the opposite side from the 'holding' side of the pattern. On initial arrival at the fix the pilot should turn on to a track 30° away from the outbound track, inside the pattern. After 1 minute (remaining inside the holding pattern), he should turn on to the inbound track to return to the fix. This is often termed a 'teardrop' join or sector 2 join.

If the arrival track is from the 'holding side' but within 110° of the outbound track (the area in Figure 7.7 marked 'sector 1') the aircraft must be flown over the fix and then along the holding leg in the reverse direction. After one minute in that direction the aircraft should turn inside the pattern through about 240° before reversing the turn on to the inbound leg. This is often referred to as a 'parallel' join or sector 1 join.

If arriving at the fix from a track requiring you to calculate the correct sector join, use a pencil or finger on the face of the instrument to represent the inbound track, and visualise the pattern superimposed. If using a RBI, you must calculate the relative bearing of that inbound track first.

The Horizontal Situation Indicator (HSI)

Many aircraft have both types of VOR display instrument fitted to navigate along VOR routes. The RMI gives a general picture of the aircraft's position, and the OBS allows easy tracking to or from a station. The HSI is a single very useful instrument which combines the two original instruments in one display for VOR or ILS (see Chapter 8) use. Figure 7.9 represents the face of a typical HSI.

Figure 7.9   HSI.

The heading pointer is at the top of the instrument, as in the RMI. It is directly fed from the aircraft's heading sensor system, as covered in the section on compasses in the volume on *Navigation* in this series. There is one RMI needle for one of the VOR inputs, of which only the ends are shown, so it is important to identify the two ends. The end with the arrowhead shows the QDM to the station (285° on Figure 7.9), and the one without marks the radial.

The part inside the compass card is the course deviation indicator (CDI) of an OBS for the other input, but with a major change. It is not oriented with the top of the instrument; the indication is presented in direct relation to the intended radial which the pilot has selected on his 'course' marker or 'bug' (CRS). As the aircraft turns, so does the CDI. The course bug is used to mark the track which the aircraft should fly. If flying away from the station, the bug should be placed on the desired radial (QDR), but if flying towards the station, the bug should be on the reciprocal of the actual radial (QDM). The CDI will therefore be lined up with the bug, and a pilot can visualise his position in relation to that track. The TO and FROM markers are also related to the course bug, showing 'TO' in Figure 7.9. As in the case of the OBS, the CDI on an HSI is marked with dots to indicate the angular distance from the desired track. When the pointer is aligned with the furthest dot, the aircraft is 10° away from the radial.

The pilot will normally use the 'heading' bug (HDG) to indicate the heading he has calculated to allow for drift, or the next heading he intends to fly, as 345° in Figure 7.9. The left side of the instrument shows the ILS glideslope indication (see Chapter 8). A warning flag, often marked 'NAV', indicates the lack of a safe usable VOR signal. Other flags may refer to other inputs.

Electronic Flight Information System (EFIS) display

Modern public transport aircraft use electronics to display the information to the pilot. One of the display modes represents a normal HSI, and other modes expand the part of the HSI display around the aircraft's track. Such an expanded mode shows the pattern of flight over a representation of the earth below the aircraft, as in Chapter 16, at Figure 16.4.

Digital displays

It is possible for a VOR radial to be displayed digitally, although this is difficult to interpret. Some hand-held communications radios have a VOR facility, which may include a digital display. A more common hand-held display would be a digital OBS numeric display with a CDI bar and dots.

Summary

RMI	• 2 inputs
	• Use for ADF/VOR or VOR/VOR
	• Tail of needle indicates radial (QDR)
	• Head of needle indicates QDM
	• Thin needle No 1
	• Fat needle No 2

> **RMI advantages**
> - QDM/QDR are indicated continuously and read off directly
> - 2 stations give instantaneous fixes
> - Useful guide when initially joining a radial for VOR homing
> - The indicator itself can be used for homing
> - The magnetic heading is shown together with QDMs
> - Relative bearings may be assessed
>
> **HSI**
> - 2 inputs
> - Use for VOR/VOR or VOR/ILS
> - Course pointer gives OBS readout from one input
> - Other needle is RMI readout from other input
> - OBI and TO/FROM lined up with course pointer

Sample questions

1. An aircraft approaches a VOR station on a track of 030°, and is instructed to hold at the fix. The holding pattern is right-hand, with a holding leg of 180°. Wind is 240°/20 kts, and TAS is 240 kts.

   a. What form of join should the pilot select?
   b. What will be the drift on the inbound leg?
   c. What should be the heading on the outbound leg?
   d. How long should the pilot maintain the outbound heading?

2. Use the RMI pictured at Figure 7.2 to answer this question.

   a. What is the aircraft heading?
   b. With reference to the VOR station selected on the No 1 needle, what is the QDM?
   c. With reference to the VOR station selected on the No 1 needle, what is the QDR?
   d. What is the relative bearing to the NDB selected on the No 2 needle?

3. A procedure turn involves an original track, a turn away for a certain period, and a final track. If the original track is 090°, what is the correct procedure for that procedure turn?

   a. Turn on to 060° for 1 min, then reverse on to 270°
   b. Turn on to 045° for 1 min, then reverse on to 270°
   c. Turn on to 060° for 75 s, then reverse on to 270°
   d. Turn on to 045° for 75 s, then reverse on to 270°

Use the following information to answer questions 4 and 5. ATC have instructed you to make a left-hand hold at an NDB, with a holding leg of 210° at FL 140.

4. You approach the NDB on a magnetic heading of 283° with a magnetic variation of 6°W. What type of join should you fly, and what is your maximum permitted indicated airspeed?

   a. A direct join, not above 230 kts
   b. A parallel join, not above 230 kts
   c. A direct join, not above 240 kts
   d. A teardrop join, not above 240 kts

5. The wind is 270° at 30 kts. Assuming your TAS is 180 kts, what heading should you adopt on the outbound leg, and for how long?

   a. 001° for 45 s
   b. 020° for 45 s
   c. 001° for 1 min
   d. 020° for 1 min

6. If the RMI indicates a VOR radial of 150°, what is the bearing of the aircraft from the station, if the variation at the aircraft is 4°W and at the station 2°W?

   a. 148°(M)
   b. 150°(M)
   c. 152°(M)
   d. 154°(M)

7. An aircraft receives a radial from a VOR at long range. At the same time it measures an ADF bearing from an NDB which is co-located with the VOR. Which of the following will be most correct on the RMI?

   a. The relative bearing of the NDB from the aircraft
   b. The magnetic bearing of the VOR from the aircraft
   c. The magnetic bearing of the aircraft from the NDB
   d. The relative bearing of the VOR from the aircraft

Chapter 8
Instrument Landing System and Markers

Introduction

The Instrument Landing System (ILS) was developed at the end of the Second World War, and allows pilots to position their aircraft in three dimensions during their approach to land. It is possible, given sufficient accuracy of installation and airborne equipment, to guide the aircraft to the runway surface itself, and that was indeed the original concept.

The three components of the system are the localiser equipment, the glidepath equipment, and the marker beacons, each with their associated monitor systems, remote control and indicator equipment. The localiser and glidepath signal emissions are designated A8W.

Basic principle

In each plane, vertical and horizontal, a pair of antennas transmit two directional beams, directed close to either side of the ideal approach path. Each beam can be thought of as activating an electromagnet to either side of a centrally biased instrument needle. When the field strengths of the beams are equal, the bias will hold the needle in the centre of the instrument. When the aircraft is to one side of the ideal path, one field strength will be stronger than the other. The needle will be deflected to the appropriate side by an amount proportional to the difference in strengths, showing the angular displacement from the ideal path. By keeping the indication central, the pilot or autopilot can maintain the ideal approach path in three dimensions down to the runway, or at least to a 'decision height' from which he can either land visually or divert if he cannot see the runway.

The actual transmissions use a different frequency for each plane. All signals are horizontally polarised. Each antenna in the pair modulates the carrier wave at a different frequency (90 Hz and 150 Hz), and it is the 'difference in depth of modulation' (DDM) which moves the needle. The signal strengths are arranged so that, when close to the intended path, the

DDM varies linearly as the aircraft moves away from the path, allowing a linear presentation on the instrument.

Localiser signals

The transmission frequency for guidance in the horizontal plane, the 'localiser' frequency, is in the VHF band, between 108 and 111.975 MHz, using the odd first decimals, such as 110.50 or 109.35. (VOR beacons may use the even first decimals.) There are 40 channels for current use, at 50 kHz spacing. A further carrier wave is also transmitted, and this can carry other information, such as an A2A identification code, airfield information, or even ATC instructions (except in category III installations).

Glidepath signals

The frequency used in the vertical plane, or 'glidepath' frequency, is in the UHF band (officially between 328.6 and 335.4 MHz, but actually between 329.30 and 335.00), at 150 kHz spacing. Every glidepath frequency is paired with a discrete localiser frequency, so that the equipment has only to be tuned to the localiser frequency for the glidepath frequency to be automatically selected. ICAO Annex 10 permits systems using two frequencies for either localiser or glidepath, but these are unusual.

The ideal polar diagram of a localiser signal is shown in Figure 8.1. The 90 Hz modulation is applied to the signal on the left side of the centreline, and the 150 Hz modulation to the right. The additional carrier wave is omitted for simplicity. An ideal glidepath radiation pattern is shown in Figure 8.2. The 90 Hz modulation is applied to the signal above the glideslope, and the 150 Hz below. The beams are arranged to give the desired glidepath angle. In most cases this would be the 'nominal' glidepath angle of 3° (a 4.9% slope) above the horizontal, but local conditions may require a different angle.

To confirm the aircraft's range from the runway, marker beacons are provided under the ideal approach path which transmit an independent

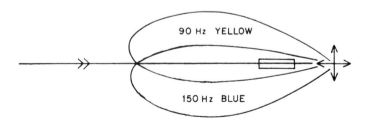

Figure 8.1 Localiser signals.

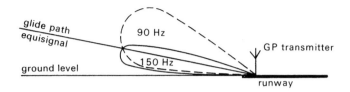

Figure 8.2   Glidepath signals.

signal on a frequency of 75 MHz. When a marker signal is received, a light appears on the instrument panel, and an audio tone is heard.

Ground equipment

The shape and direction of the localiser beams is controlled by the localiser aerial. In accordance with ICAO Annex 10, this must be placed beyond the end of the runway, at a safe distance to prevent it becoming an obstruction, in line with the runway. In some situations, the aerial may have to be placed to one side, which means it can only be used as a category I ILS (see below). The localiser centreline will then be at an angle from the runway centreline, and the installation is called an 'offset' ILS. The localiser centreline of an offset ILS will cross the runway extended centreline at the same range from the threshold as an approaching aircraft would reach its decision height. If the offset is more than 2°, the ILS cannot be used as a 'precision approach aid'. It is regarded as an 'airfield approach aid' only.

Markers

There may be three marker antennas, although two is the minimum, sometimes in conjunction with an NDB. The 'outer marker' is positioned at a range from touchdown sufficient to give height, distance and equipment functioning checks, usually just after the pilot descends on the final approach (Annex 10 recommends 3.9 nm from touchdown, but certainly between 3.5 and 6 nm). A 'middle marker' is placed (ideally 1050 m from touchdown), to indicate that visual references should be available, in other words around category I decision height. The optional 'inner marker', if fitted, is placed just short of the threshold itself, in the area where an aircraft would be at a category II decision height.

Aircraft have to be able to receive the marker signals on the glideslope for enough time to identify them and make calculations, so the beam is shaped to produce a signal over specific distances at the glideslope. Ideally, the outer marker should be 600 m wide, the middle marker 300 m, and the inner marker 150 m. The beam is broad enough to allow an aircraft within 2.5° of the centreline to receive it. Representative horizontal polar diagrams of marker beacons (including en-route markers described later) are shown at Figure 8.3, with the aircraft's intended direction of travel being from right to left.

inner   middle   outer   fan   Z

Figure 8.3   Marker horizontal emission patterns.

ILS reference point

The antenna for the glideslope is positioned alongside the runway, in such a position that the glideslope passes over the threshold at a height of 50 ft. This point in space is called the 'ILS reference point'. Figure 8.4 shows a plan view of a runway with the position of the ILS antennas marked.

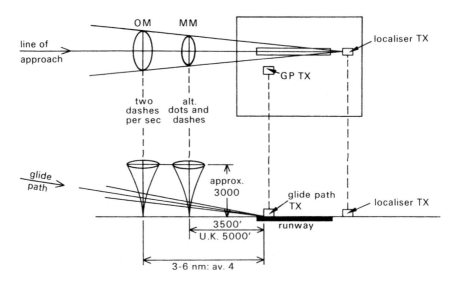

Figure 8.4   ILS ground antenna positioning.

DME

Many ILS systems have a DME (Distance Measuring Equipment, which is covered in Chapter 14) frequency paired with them. In that case, the DME is intended to replace the markers, to give range indications from touchdown during the final approach. The output from such a DME is electronically adjusted so that the received ranges are only correct when the aircraft is on the ILS centreline in the direction of approach.

Monitoring

Each component of the ILS ground equipment contains its own monitoring system, which will switch off either the carrier waves or the modulation signals if they are outside the required parameters, or indeed if the monitoring system fails. As an exception, if the installation is category III, signals may be continued in certain cases but the guidance will not be adequate for category III, so the installation would be downgraded to category I in those cases, and the modulation of the carrier wave would contain that information.

Aircraft equipment

OBS display

The ILS signals are often displayed on an OBS with an extra horizontal bar and dots arranged vertically, similar to those of the CDI, as in Figure 8.5. This 'glidepath pointer' indicates a maximum deviation of 0.7° from the ideal, therefore each dot indicates a deviation of 0.14°. When receiving an ILS localiser signal, the CDI itself indicates a maximum deviation of 2.5° from the centreline, much less than when using a VOR signal (10°).

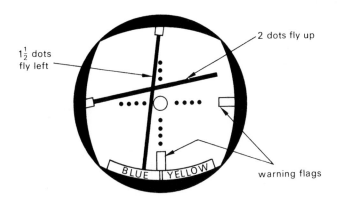

Figure 8.5   ILS indicator.

Warning flags are fitted to indicate lack of usable signal from either of the two paired frequencies or the modulating frequencies. In the event of a glideslope signal not being available, it is possible to use the localiser only for a non-precision approach.

Flight director (zero reader)

A specialised instrument was devised for ILS signals, and can also be used for VOR navigation. A computer takes track and altitude information from the aircraft's air data computer and Doppler radar or other source, as described later in this book, as well as the ILS signals and course pointer. The computer calculates the most efficient path for the aircraft to follow to put the aircraft on the centreline and glidepath, and presents the information to the pilot as a point to which he should point the aircraft. This may be two crossed pointers in a separate instrument (sometimes called a zero reader), which require the pilot to fly the aircraft to keep the crossbars central by varying pitch and bank attitude. A zero reader is shown in Figure 8.6. A more common display has markers ('tramlines') on the aircraft's attitude indicator on which the pilot should place the horizon bar, as in Figure 8.7. The ADI in Figure 8.7 also indicates the raw ILS centreline information along the bottom, and glideslope information along the right-hand side of the display.

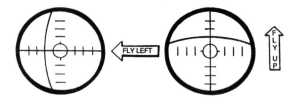

Figure 8.6   Zero reader flight director.

Fig. 8.7   ADI (courtesy Meggitt Avionics).

HSI presentation

The centre of the HSI contains the CDI, which, as mentioned in Chapter 7, turns with the compass card to indicate orientation. It is used in the same manner as the CDI of the OBS. However, the glideslope is frequently indicated by marks at the side of the instrument, with usually a diamond-shaped lozenge to indicate the aircraft's position in relation to it, as in Figure 8.8. Again, warning flags indicate lack of usable signals.

Figure 8.8 HSI.

Marker indicators

Lights on the instrument panel illuminate when the aircraft passes within the marker signals, and audio notes can be heard. Passing the outer marker, the pilot sees a blue light and hears a series of low pitched dashes (2 per second at 400 Hz). Passing the middle marker, an amber light illuminates, and the audio signal sounds dots and dashes (at 1300 Hz). Passing the inner marker, if installed, a white light illuminates and a series of high pitched dots can be heard (6 per second at 3000 Hz). Figure 8.9 shows a marker

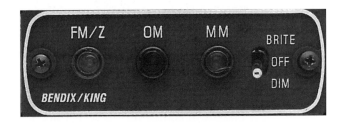

Figure 8.9 Marker beacon display unit (courtesy Honeywell).

beacon display panel. That marked FM/Z is designed for fan or Z marker signals (see later), but will display inner marker signals also.

Aerials

Being in the VHF band, the localiser signals are often received by whip antennas on either side of the tail, which may also be used for VOR signal reception. The UHF glideslope signals can be received by slot antennas in the wing leading edge. The lower frequency marker signals require a larger antenna, which is usually mounted below the fuselage.

Control unit

A standard communications/navigation box, a modern example of which is illustrated in Figure 8.10, is often fitted in light aircraft. When selected to an ILS frequency, it allows ILS information to be automatically displayed on the flight instruments. The illustrated unit controls electronic screens to display the ILS commands, although it includes a basic OBI. Larger aircraft may have individual selectors.

Figure 8.10   Comm/Nav/ILS control unit (courtesy Honeywell).

Use of the ILS

The pilot should first tune the required ILS localiser frequency and identify the station from the audio signal. The course pointer or OBS should be set to the localiser centreline magnetic track as published, although some displays may not require this. The localiser signal will then give an indication of the aircraft's position in relation to the runway centreline extended at either end, but no other information. Some other aid is normally required to position the aircraft in a suitable position to make its approach. However, if the track to the aerodrome is known to within 180°, it is possible to use the ILS indications themselves for an approach, although in a rather inefficient manner.

Initial approach

If the rough track to the aerodrome is known, the pilot can intercept the localiser some distance along it, in a similar fashion to a VOR radial intercept. The sequence of markers will confirm the position of the airfield, and, after a procedure turn back along the localiser, the aircraft can descend safely outbound along the localiser in the reverse direction. On a simple display, the localiser signals will provide reverse directions at this stage. A further procedure turn after the outer marker can bring the aircraft back on to the localiser beam, where glideslope information may be followed as detailed below.

Such an approach procedure would be time and airspace consuming. Much more commonly, a pilot will use another aid, often an NDB on the airfield or under the approach path, to bring the aircraft directly to a safe area. There it can be descended to a safe altitude during a published procedure leading the aircraft towards the localiser centreline.

Final approach

Once on the localiser beam, and tracking towards the runway using a suitable heading compensated for drift, the aircraft must fly towards the glideslope from below. The outer marker will confirm the aircraft's position close to the descent point. When the glideslope pointer indicates that the aircraft is on the slope, the pilot adjusts power or configuration to maintain a suitable rate of descent to keep him on the slope. The rate of descent depends on the aircraft's airspeed, the glideslope angle, and the wind component.

The rate of descent required can be calculated by a formula using the 1 in 60 rule. For a 3° slope, the aircraft must descend 300 ft every mile. For a 2.5° slope, the descent rate must be 250 ft per mile, and similarly for a 4° slope, 400 ft every mile. By multiplying that figure by the groundspeed in miles per minute, a datum rate of descent can be calculated. For example, for an aircraft travelling at 140 kts TAS with a 20 kt headwind (120 kt groundspeed, or 2 miles per minute), to follow a 3° slope the pilot should adopt a rate of descent of $2 \times 3 = 600$ ft/min. Calculations of this type are typical examination questions. (For 3° slopes only, you may notice that a simple way of achieving the answer is to multiply the groundspeed by 5.)

As the aircraft descends, the pilot must adjust heading and rate of descent to maintain the centreline and glideslope pointers in the centre. During the descent, the wind will back and decrease, as explained in *Ground Studies for Pilots: Meteorology*, and gusts will affect the flight path. Adjustments must be made to take account of those. Because the instrument indicates angular displacement, as the aircraft approaches the threshold the pointers appear much more sensitive, and smaller corrections are needed for the same needle deflection.

Pilot monitoring

A pilot should not rely implicitly on any navigation aid. During the final approach, he should check that his height above touchdown at known ranges is that which he expects. For a 3° approach, the aircraft at four miles from touchdown should be at a height of 1200 ft above touchdown, and most approach 'plates' will publish the expected heights or altitudes when passing the markers or whole miles of DME ranges. Examination questions may ask candidates to calculate the expected height at a certain range from the runway threshold. In such cases, it must be remembered that the ILS reference point is 50 ft above that threshold, so that 50 ft should be added.

Back beam approaches

The shape of the localiser emission pattern provides information along the runway centreline in both directions. It is possible (but not permitted in the UK) to follow the localiser beam towards the runway reciprocal to the runway along which the ILS is positioned, and some procedures employ this. On a simple display, the directions have to be followed in reverse, although flight directors may have a switch which allows normal indications on back beam approaches (or outbound along the localiser). Back beam approaches are non-precision.

Operational categories

ICAO categorises ILS equipment and procedures. The operational objectives (what they are designed for) for these categories are as follows:

- Category (Cat) I – to allow aircraft to approach a runway in a position to land with a decision height of 60 m (200 ft) above the runway, and a visibility of not less than 800 m or runway visual range (RVR) of not less than 550 m.
- Cat II – to allow aircraft to approach a runway in a position to land with a decision height of between 60 m and 30 m (100 ft) and a RVR of not less than 350 m
- Cat IIIa – to allow aircraft to approach a runway in a position to land with a decision height of 30 m or less (or no decision height) and RVR of not less than 200 m.
- Cat IIIb – to allow aircraft to reach a position to land with a decision height of 15 m (50 ft) or less, and RVR between 200 m and 50 m.

- Cat IIIc – to allow aircraft to land on the runway with no cloudbase or visibility restrictions.

The actual definitions in ICAO Annex 10 are as follows:

- Cat I – An ILS which provides guidance from the coverage limit to the point at which the localiser course line intersects the glide path at a height of 60 m or less above the horizontal plane containing the threshold.
- Cat II – An ILS which provides guidance from the coverage limit to the point at which the localiser course line intersects the glide path at a height of 15 m or less above the horizontal plane containing the threshold.
- Cat III – An ILS which, with the aid of ancillary equipment where necessary, provides guidance from the coverage limit to, and along, the surface of the runway.

It may be convenient to think of the categories as:

- I allowing a manual approach to a decision height for visual landing,
- II allowing an autopilot approach with manual visual landing, and
- III allowing automatic landings only;
- IIIa giving no directions on the runway itself,
- IIIb requiring visual monitoring of the direction on the runway, and
- IIIc allowing the aircraft to stop using the ILS signals.

Possible problems

False glideslope

Because of the emission pattern of the glideslope antenna, the 150 Hz signal will be received above the intended glideslope. This will give false, indeed reverse, indications, and will occur at an angle of twice the nominal angle, for example at 6° for a 3° standard glideslope. To prevent the possibility of following the false glideslope, aircraft must always approach the glideslope from below, and procedures are designed to achieve this. Pilots should also be aware of the direction of movement of the glideslope needle before they use it.

Signal reflection (beam bending)

The localiser signal will diffract (bend) around many obstructions, although reflections can affect the relative strength of the modulated signals. At close range (below Cat I decision heights), such effects from aircraft or vehicles close to the localiser beam, for example at the holding point next to the threshold, can cause an apparent 'bending' of the localiser beam. Aerodrome operators attempt to minimise this by providing a separate holding point while Cat II or III operations are in force.

Limitations

System limitations

ICAO lays down a volume of airspace within which the system must provide accurate indications. Although correct indications will be received in other places, aircraft should not use the system outside that volume, as detailed below.

The localiser beam must be accurate from airfield level up to a vertical angle of 7°, within 10° of the centreline out to 25 nm from the intended point of touchdown. It must be accurate within 35° of the centreline out to 17 nm, as shown in plan view in Figure 8.11. This may not always be practical, in which case the systems are designated 'steep'. For these systems, the horizontal angle must be 10° out to 18 nm, and 35° out to 10 nm.

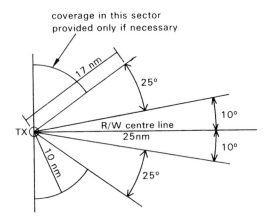

Figure 8.11 Standard localiser beam 'footprint'.

The glidepath signal must be accurate over a vertical angle of between 45% and 175% of the ideal approach slope (for a 3° slope that means from 1.35° up to 5.25°) out to 10 nm and within a horizontal angle of 8° of the centreline, as shown in Figure 8.12.

Markers must retain their emission pattern shape up to a height of 3000 ft above the airfield level. In fact, all measurements are made from the 'reference point' which is the point in space above the runway threshold where an aircraft's wheels are intended to pass.

DMEs paired with ILS installations are protected from interference from

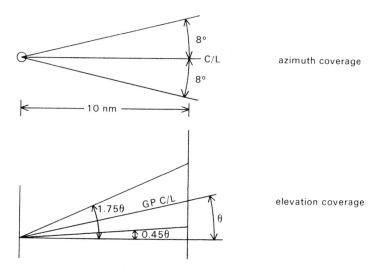

Figure 8.12   Glidepath beam 'footprint'.

other DMEs only within the 'footprint' or area of accurate localiser signals, up to a height of 25 000 ft.

Operating limitations

Pilots must not rely on an ILS system outside the ICAO 'cleared volumes'. Nor should they use the system outside aerodrome opening hours. In addition, they must not commence descent on the glideslope unless they are within 2.5° of the localiser and established on it. If, having commenced descent on the glideslope, the indication reaches half of its full 'fly up' indication, national regulations require that the crew carry out a go-around immediately.

If the ILS is unserviceable or undergoing maintenance, the identification modulation will be removed or replaced by a continuous tone. In these circumstances, the ILS must not be used. For that reason, it is advisable to continuously monitor the identification signal during the approach.

En-route markers

Marker beacons are sometimes used along airways. These transmit an uninterrupted horizontally polarised 3000 Hz signal on 75 MHz, and are received on ILS equipment as a constant audio note and steady white light. This is often the same light that provides indication of the inner marker on an ILS approach, as in the unit illustrated in Figure 8.8. En-route markers may be modulated with dots and dashes for identification.

Fan marker

A fan marker, the usual form of en-route marker, radiates a pattern vertically upwards, ideally in the form of a 'rectangular parallelpiped', or tube with a constant rectangular cross-section, orientated with the wider part across the flight path of an aircraft intended to receive it.

Z marker

Some en-route markers transmit signals vertically upwards in the form of a cylinder with a circular cross-section. These are called 'Z markers'.

ILS summary

Principles	Difference in depth of modulation of 2 lobes fed to needle on indicator
Frequencies	Localiser VHF (108 to 112 MHz, odd first decimals), glideslope UHF
	90 Hz on left and high, 150 Hz on right and below
Emissions	A8W
Localiser	25 nm, $\pm\,10°$ along centreline; 17 nm, $\pm\,35°$. (18 nm and 10 nm for steep system)
	Vertical cover 7°
Glideslope	$0.45 \times$ G/S angle to $1.75 \times$ G/S angle, horizontal $\pm\,8°$ out to 10 nm
Indicator	Full deflection localiser = 2.5°, glideslope = 0.7°
	1/2 'Fly Up' = Initiate Climb!
Markers	Fan beams, up to 3000 ft
	75 MHz – Outer – blue light and low pitched dashes
	Middle – amber light and mid pitch both
	Inner – white light and high pitch dots
Reference	Height of glideslope centre above threshold (50 ft)
Operational categories	1 = DH not below 60 m, RVR 550 m
	2 = DH not below 30 m, RVR 350 m
	3 = DH below 30 m if any, appropriate RVR
Calculations	Heights on glideslope (add 50 ft for reference if needed)
	Rates of descent (G/S × for 3° slope)
DME	When paired with ILS, DME fudged to give range from touchdown only along centreline
Offset	If localiser offset by more than 2°, ILS becomes non-precision aid

False glidepath	Found sometimes at 2 × normal glideslope. Approach from below
Back beam	Ignore any such indications in the UK
Low visibility procedures	Holding well short of runway to avoid interference
	Moving large objects cause greatest interference
En-route markers	White light. Fan markers rectangular, Z markers circular cross-section

Sample questions

1. Where would you expect to find a false glidescope on an ILS?

   a. On the localiser, below the correct glideslope
   b. On the back beam, below the correct glideslope
   c. On the localiser, above the correct glideslope
   d. On the back beam, below the correct glideslope

2. Which frequency bands are transmitted by ILS equipment?

   a. Glideslope on VHF, localiser on VHF
   b. Glideslope on VHF, localiser on UHF
   c. Glideslope on UHF, localiser on VHF
   d. Glideslope on UHF, localiser on UHF

3. A category I ILS allows approaches with the following criteria:

   a. Decision height below 30 m, an appropriate RVR?
   b. Decision height not below 30 m, RVR 350 m?
   c. Decision height not below 60 m, RVR 350 m?
   d. Decision height not below 60 m, RVR 550 m?

4. The outer marker of an ILS system emits:

   a. Blue flashes synchronised with audio dashes?
   b. White flashes synchronised with audio dashes?
   c. Blue flashes synchronised with audio dots?
   d. White flashes synchronised with audio dots?

5. What rate of descent should an aircraft use on a 3.5° ILS glideslope if it has a groundspeed of 120 kts?

   a. 500 ft per min

   b. 600 ft per min
   c. 700 ft per min
   d. 800 ft per min

6. The frequency band 108 MHz to 112 MHz contains:

   a. VOR signals only?
   b. ILS signals only?
   c. VOR and ILS signals?
   d. VOR and VHF communications signals?

7. An ILS glideslope signal may be received within the following footprint:

   a. 8° either side of the centreline, out to 25 nm?
   b. 8° either side of the centreline, out to 10 nm?
   c. 10° either side of the centreline, out to 25 nm?
   d. 10° either side of the centreline, out to 10 nm?

8. An ILS glideslope indicator measures:

   a. The difference in strength between 2 signals of 90 MHz and 120 MHz?
   b. The difference in phase between 2 signals in the UHF band?
   c. The difference in modulation depth of 2 signals?
   d   The difference in phase between 2 signals in the VHF band?

9. What height should an aircraft on a 2.5° glideslope be, at a range of 6 nm from touchdown, given an ILS reference of 50 ft?

   a. 1350 ft
   b. 1550 ft
   c. 1750 ft
   d. 1950 ft

10. What form of emission does an ILS signal use?

   a. A2A
   b. A3E
   c. A8W
   d. J3E

11. A category III ILS allows approaches with the following criteria:

   a. Decision height below 30 m, an appropriate RVR?
   b. Decision height not below 30 m, RVR 350 m?
   c. Decision height not below 60 m, RVR 350 m?
   d. Decision height not below 60 m, RVR 550 m?

12. What rate of descent should an aircraft use on a 2.5° ILS glideslope if it has a groundspeed of 120 kts?

    a. 500 ft per min
    b. 600 ft per min
    c. 700 ft per min
    d. 800 ft per min

13. If a DME is paired with an ILS, where will the indicated ranges be correct?

    a. In all directions, up to 20 nm
    b. Within the localiser catchment area
    c. In all directions, up to 10 nm
    d. Along the runway centreline in both directions, up to 20 nm

Chapter 9
Basic Radar

Introduction

The acronym RADAR comes from 'RAdio Detection And Ranging', which suggests its military origins, but also gives an insight into the original theory. We have already seen that a radio wave can be reflected by objects, just as visible light waves can be reflected. Some of the waves will be reflected directly back towards the transmitter. A sensitive receiver tuned to the transmitted frequency can detect the reflected waves, and therefore the object reflecting them. If the waves are transmitted in short bursts or pulses then it is possible to measure the time between the transmission of the original signal and the reception of the reflection. Because electromagnetic waves travel at a virtually constant speed through a particular medium, such as air, the time from transmission to reception is proportional to the distance the signal has travelled and therefore the range of the object from the transmitter. This was the earliest form of radar use.

The reflectivity of an object depends on the shape it presents to the signal beam, like a mirror reflecting light. However, it also depends to a considerable extent on its size relative to the wavelength of the radio signal. Ships, for example, are large and can reflect quite long wavelength (hence low frequency) signals. Raindrops, however, only reflect very short wavelength (hence very high frequency) signals.

By sending a narrow, directional, beam of waves it became easy to determine whether the reflecting object (target) was contained in the beam. This required an aerial which could send a narrow beam, so a parabolic reflector antenna was used in the transmitter, as in Figure 9.1. To determine the actual direction of the target, the beam had to be moved around, so the antenna was turned mechanically. Because the transmitter only sent short pulses, and then the equipment had to wait for the reply, the same antenna could be used for the receiver, saving materials, weight and cost.

The time base

A basic reflected radar signal can be displayed on a screen with a phosphorescent (glowing) coating, as explained in Chapter 11. The width of the

Figure 9.1   Parabolic radar antenna.

screen represents the distance out to which the operator wishes to detect a target. A stream of electrons pointing at the screen is made to move in a straight line along it at a speed proportional to the speed of radio waves across that distance. It therefore simulates the pulse of radar energy transmitted outwards. After reaching the end of that distance (the maximum range), the electron stream is brought back to the beginning again for the next pulse (fly-back). A reflected signal from a target is engineered to make a mark on this timebase by deflecting the electron beam at 90° to it, and an operator can see the target's range by comparing it with fixed marks representing ranges from the transmitter. A simple timebase is shown in Figure 9.2.

Direction finding

A basic radar only shows ranges, but once a target is detected it is possible to move the antenna until the signal is no longer received, which will give an indication of the target's bearing from the receiver. Moving the timebase with the antenna will show that direction, as the target will 'paint' when it is within the polar diagram of the antenna, and will slowly disappear (fade) when no reflected signal is received.

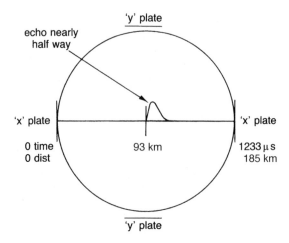

Figure 9.2   Radar echo on a simple timebase.

The plan position indicator (PPI)

The antenna can be moved horizontally, in a circular 'sweep'. As the antenna rotates slowly, the beam of radiation also rotates. While the target is being 'illuminated' by the beam, the reflection will be received. There are many pulses sent during the period when the target is being illuminated, and each reflected signal will be received at the same time after each transmission.

The display for the radar operator is usually a cathode ray tube, covered in detail elsewhere. In essence, the timebase rotates around the centre of the tube, and each received reflection 'paints' in the same position on the display, for as long as the signal is being reflected. When the reflection is no longer received, there is no more 'painting', and the target 'fades'. The fade is slow enough to show the target faintly long after the next set of paints have arrived with the rotating timebase. A moving target will leave a trail of these fading paints to indicate its past movement as well as its current position. A PPI, an example of which is shown in Figure 9.3, can display many targets.

Sector scanning

A moving target will change its position between sweeps. If it is desired to provide further information on the movement of a particular target, a separate antenna can be moved back and forth in a sector scan (reciprocating) so that the target is illuminated for a much longer period. If the scan is vertical, the display can measure height. Traditional height finding radars are directed towards their target by operators using a PPI display initially.

Precision approach radars use the sector scanning technique with two antennas to monitor the target in both the horizontal and the vertical.

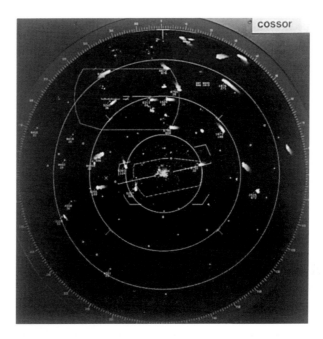

Figure 9.3   PPI display.

Terminology

A representation of pulse transmission is shown in Figure 9.4.

Pulse width

The length of time that a radar pulse is transmitted is called the pulse width. It is usually measured in microseconds.

Pulse length

The length in metres occupied in space by a transmitted pulse. The expression is also sometimes used as an alternative to 'pulse width' and measured in microseconds.

Pulse repetition interval (pri)

The time between the start of consecutive radar pulses is called the pulse repetition interval, or pulse repetition period (prp). Again this is usually measured in microseconds.

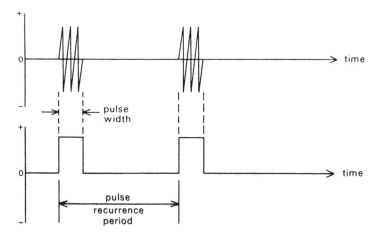

Figure 9.4 Radar pulses.

Pulse repetition frequency (prf)

The number of pulses per second (pps) is called the pulse repetition frequency, or sometimes the 'pulse rate'. Mathematically one second divided by the pri equals the prf, so they can be regarded as the reciprocal of each other.

Scan rate

The beam from a rotating antenna will illuminate a target each time it rotates, so its scan frequency is the same as the number of antenna rotations in a given time, usually every minute. This is often referred to as the 'antenna rpm'. The scan rate is of course different for a reciprocating aerial, being measured as twice every double sweep, reciprocation, or 'nod'.

Beam width

The shaped antenna produces a polar diagram in a particular direction, whose ideal pattern is shown in Figure 9.5. The maximum strength is in the direction the antenna is pointing, and the 'beam width' is measured as the angle between the two points where the signal strength is half that maximum. A wide antenna shape produces a narrow beamwidth in that direction, so the aerial at Figure 9.1 would have a narrow beamwidth in the horizontal plane and a broader beam width in the vertical plane.

This idealised pattern does not exist, however. The pattern has 'side lobes' of energy which alter the polar diagram as shown in Chapter 17 at Figure 17.2.

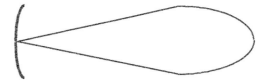

Figure 9.5   Parabolic antenna ideal polar diagram.

Radar parameters

Maximum range

Several factors affect the maximum range available from a radar set. One of the major ones is the time the receiver waits for the reflected signal to return before sending the next pulse (the pri). This maximum range can be calculated theoretically by using the formula $R = Cpri/2$, where R is the maximum range, and C is the speed of radio waves in air, usually accepted as 3×10^9 m/s. (Of course, the result includes division by 2 because the signal must travel out and back before it is received.) The formula can also be written in its more common form of $R = C/2prf$.

Another factor is the power available, which is covered later. Because the signal is reflected, and the power is lost in both directions, the power at the receiver and therefore the range available is proportional to the fourth root of the transmission power, or $R \propto \sqrt[4]{power}$

Minimum range

While the transmitter is operating, the antenna cannot receive a reflected signal. The time for the transmission is the pulse width, hence the minimum theoretical range can be calculated using the formula $r = C/pw$ where pw is the pulse width in seconds. This is more easily calculated as $r = 150/PW$ where PW is the pulse width in microseconds, the common unit. If one is fortunate enough to know the pulse length in distance units, however, the minimum theoretical range is simply half that pulse length.

Range discrimination

Where more than one target is illuminated by a particular transmitted wave, the targets will merge if the reflection from the closer target is still being received at the antenna when the reflection from the further target arrives. In other words, if the targets are closer together than the minimum theoretical range calculated as above, there will be no discrimination between the targets and confusion may result.

Angular discrimination

The same theory applies to angular separation between two targets. As the antenna rotates, if two targets are being reflected by the same beam the receiver cannot discriminate between them. Discrimination is possible if the targets are further apart than the beam width, so the angular (azimuth) discrimination equals the beam width. The actual distance can be calculated mathematically, or in practice by the 1 in 60 rule. Figure 9.6 shows an example of azimuth discrimination between two targets approaching the radar antenna.

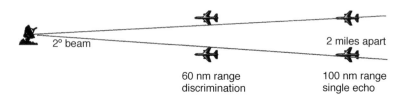

Figure 9.6 Azimuth discrimination.

Power available

The power of the signal generated at the transmitter is directed by the antenna, so the beam width is a factor in the strength of the signal reaching the target and therefore reflected. A narrow beam width gives more power, so the power is inversely proportional to the beam width (in both directions, vertical and horizontal). The pulse width is another factor, because the longer the energy generated at the transmitter can be transmitted the more is available for reflection. The reflected power is therefore directly proportional to the pulse width.

Use of radar

The use to which a radar is to be put determines the design parameters of the transmitter and antenna. As already stated, the size of the target often determines the signal frequency. For example, a weather radar requires a wavelength of about 3 cm to reflect large raindrops while ignoring small ones, hence a frequency of about 10 GHz. Similar but usually slightly longer wavelengths (to avoid saturation by rain returns) at corresponding frequencies can resolve detail from larger objects, such as shapes of aircraft on an airfield ground movement radar.

If maximum range is required, the power must be concentrated with a long pulse width (which has the disadvantage of reducing range resolution) and narrow beam width. The pri must also be long, which coupled with the narrow beam width requires a low rotation frequency (slow sweep) which has the disadvantage of leaving a long gap between illuminations.

A radar such as a ground movement radar on an airfield does not require a long maximum range, but does require a very short minimum range. Hence such a radar will be designed with a short pulse width, and the short maximum range required will allow a short pri and hence a rapid rotation frequency.

Modern radars

Modern radars do not always require the antennas to rotate physically in order to sweep the beam. A number of transmitting or receiving antennas within a large array can be electronically switched by a computer in such a way that the resultant beam direction changes as if the aerial was actually moving. Nor do modern radars require to display the actual received signals. It is usual for a computer to convert the received signals to digital inputs which can be processed before display on a computer monitor.

Using special, very large, aerials, the ground waves of medium frequency signals can be used in extremely long range, 'over the horizon' radars. At the time of writing, these are mainly used by the military, but their use by civilian air traffic controllers is quite feasible. In certain remote parts of the world, such as the north of Siberia, they may provide the best means of controlling air traffic safely.

Summary

Principle	Pulse of energy transmitted – time for reflection proportional to range
	Directional aerial rotates to give bearing
	Echo shows on cathode ray tube – fades slowly
Definitions	Pulse width (or length) = time spent transmitting pulse (sometimes actual pulse length)
	Pulse repetition period (prp) = time between pulses
	Pulse repetition frequency (prf) = number of pulses per second
	Beam width = angular width of transmitted beam
Max range	Signal travels to and from the target at 300 million m/s (c)
	Range = $c \div \text{prf} \div 2$ (or $c \times \text{prp} \div 2$)
	Max range is proportional to $\sqrt[4]{\text{power}}$ as well as prp
	VHF formula if power and prp enough
Min range	= Pulse width $\times c \div 2$

Definition	2 targets must be 0.5 pulse width apart to be distinguished in range
	Must be beam width apart to be distinguished in azimuth
Beam width	Depends on aerial design
Power	Long pulse width = more power

Sample questions

1. A radar transmission has a pulse width of $2\,\mu s$. What is its minimum resolution distance in range?

   a. 3 km
   b. 600 m
   c. 300 m
   d. 60 m

2. What maximum detection range would you expect from a radar transmission with a prf of 1 kHz?

   a. 15 km
   b. 30 km
   c. 150 km
   d. 300 km

3. A ground radar with a frequency of 10 GHz would be used for:

   a. Terminal area control and surveillance?
   b. Airways monitoring?
   c. Precision approach radar?
   d. Airfield surface movement surveillance?

4. Azimuth resolution on a radar screen depends on the:

   a. Pulsewidth of the transmission?
   b. Frequency of the transmission?
   c. Beam width of the transmission?
   d. prf of the transmission?

5. What band does a surface movement radar on an airfield use?

   a. SHF
   b. VLF
   c. VHF
   d. HF

6. A radar antenna rotates at 15 rpm. It has a pulse width of 2 µs, and a beam width of 3°. Two aircraft are flying towards the antenna at the same speed and initial range, but separated laterally by 1 nm. At what range from the antenna will the radar first be able to distinguish between the two aircraft?

   a. 15 nm
   b. 20 nm
   c. 30 nm
   d. 45 nm

Chapter 10
Ground Radar Services

Introduction

We have already looked at the basic radar theory. However, the syllabus for commercial and ATP licences now includes not only the types of radar and their descriptions, but also the procedures laid down by ICAO for Air Traffic Control purposes. In this chapter we shall look at the ground based primary radar equipment and procedures used by air traffic controllers in their job of providing assistance to pilots.

Types of ground radars

Search radars

These are referred to by ICAO as 'primary surveillance radars' or PSRs. An en-route surveillance radar (sometimes called RSR) normally requires a range of 200 miles or more, and good range and azimuth resolution. Pulse widths have to be short, and beam width narrow, so high transmitter power is needed to provide a strong reflected signal at the receiver. The frequencies tend to be around 400 MHz.

Displays are PPI types. Long range search radars have a sweep rate of about once every 12 s. Terminal Area Surveillance Radars (TAR), used for guidance close to aerodromes and with a range of about 25 nm, scan quicker, down to once every 3 s. Controllers use search radars to give directions (radar vectors) to pilots to guide them to an aerodrome, to ensure smooth flow of traffic, and to avoid collisions or weather.

Height finding radars

The sector scanning antennas of traditional height finding radars have a characteristic 'nodding' effect. Mechanically directed to a target detected on the search radar, the display is in the shape of a vertical sector. The height of the target can be found by multiplying the measured range by the sine of the angle above the horizon and adjusting the units. Most height finding radars have been withdrawn from air traffic control systems, SSR with altitude reporting (see Chapter 17) has taken over. However, height finders, or more precisely angle finders, are used in PARs, as explained below.

Some modern primary search radars use special aerial arrays which shape the signal beam electronically in such a way that it 'sweeps' its scan area in a 'raster' pattern (similar to that described in Chapter 11). This can provide height finding simultaneously with direction finding. The cost of such equipment tends to restrict its use to military applications.

Precision approach radars (PAR)

To provide accurate indications of an aircraft's position in three dimensions when it is approaching the runway, two sweeping radars can be used. One, height finder, is used for 'glidepath' guidance, and another sweeps horizontally for 'centreline' guidance. These do not require a long maximum range (typically 10 miles is enough), but need good definition and a short minimum range. The wavelength is around 3 cm, so rain and hail can interfere considerably with the radar pictures. Few civilian aerodromes in the UK use PAR at the time of writing; it has been superseded by ILS installations, but many military aerodromes use it as the primary runway approach aid.

The displays have the correct paths (vertical and horizontal) painted on their screens, as represented in Figure 10.1 below. The screens are usually mounted as shown, one above the other, so that the controller can see both of them without moving his head. The controller gives instructions to the pilot

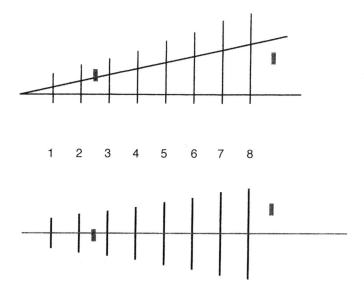

Figure 10.1 PAR display.

to keep his aircraft, which he sees as the centre of the radar return, on these paths.

The representation of the two screens in Figure 10.1 shows an aircraft at a range of 2.5 miles from touchdown (on the left of both screens), slightly above the glidepath, and slightly left of the centreline. A further aircraft is at 8.75 miles, approaching the centreline from the right, and not yet in a position to commence its descent.

Surface movement radars

Areas of large airports are often hidden from Air Traffic Control, especially in poor visibility. Many ground controllers use surface movement radars (referred to by ICAO as SMR, but also called airport surveillance detection equipment or ASDE) to assist them when directing movement on the aprons, taxiways and runways. SMR is particularly useful in ensuring runways are unobstructed when issuing take-off and landing clearances, and in guiding emergency vehicles to an incident. Their maximum range need only be a matter of two or three miles, but they require even better definition and shorter minimum range than the PAR.

Weather radars

Both Air Traffic Control and meteorological services use radars to identify weather phenomena and track radio-sonde balloons. Surveillance radars with wavelengths of about 3 cm are used to search for rainfall and thunderstorms, as in airborne weather radar (see Chapter 12). Other surveillance radars are used to track radio-sonde balloons. The movement of the target, the balloon, is an indication of the mean wind between two altitudes.

Use of surveillance radar

Before providing a service to an aircraft, the controller must first identify it from among the many targets on his screen. Initially, the pilot can be asked to state a VOR/DME position. A VDF trace will often be available on the screen to give a bearing on the pilot's transmissions. This will usually be confirmed by asking him to make certain manoeuvres, such as a turn on to a certain heading through at least 30°. If the aircraft is equipped with SSR (see Chapter 17), or the aircraft has been 'handed over' from another controller, manoeuvring is not necessary.

Once identified, the controller may provide vectors (specific headings) to guide the pilot verbally through controlled airspace or away from hazards such as weather. A radar controller who is giving vectors to an IFR flight takes responsibility for keeping a safe obstacle clearance beneath the aircraft he is vectoring, provided the plot maintains the cleared altitude. If and when the controller ceases to provide radar control, he will use the words 'resume own navigation' to indicate that the pilot once more has responsibility for navigation and obstacle clearance.

Controllers can find the position of a target aircraft by reference to the map features on the video map. It can also be found by interpolation from the range rings displayed with the picture and the angles displayed along the circumference of the screen. Modern equipment has more sophisticated means of finding exact positions. Controllers can move a 'variable range ring' until it corresponds with the blip on the screen and read the exact range from a display unit. A similar method provides bearing information by moving an 'electronic bearing line' around the screen until it lies over the blip, then reading the exact bearing from the same display unit.

ICAO lays down a minimum specification that a surveillance radar must be able to identify a small single-engined aircraft at a distance of 20 nm at 8000 ft. The position must be accurate to within $\pm 2°$, and it must be able to distinguish between two aircraft $4°$ apart in azimuth. A beam $2°$ wide and up to $30°$ in elevation provides this specification.

The scale of the radar image can be varied by the controller to show the area in which he is interested. An en-route radar can display areas with radii (displayed ranges) down to 20 nm. Radars with a wavelength about 3 cm or less are affected considerably by rain and hail, being both reflected and absorbed by precipitation. As wavelengths increase, absorption reduces rapidly, and reflection reduces slowly. Longer wavelengths continue to reduce the effects, and at a figure of around 50 cm wavelength avoid weather interference altogether, but most search radars are affected to a greater or lesser extent.

Surveillance radar approach (SRA)

For use in aerodrome approaches, the controller vectors the aircraft towards the approach path, giving instructions to descend to usually 1500 ft above the aerodrome before it arrives on the extended runway centreline. Once established on the centreline, the controller will vector the aircraft towards the final approach fix. At that point, he will instruct the pilot to commence a descent towards the threshold. On a SRA the pilot will select a rate of descent which will bring him to the runway threshold, without descending below a series of 'check heights' which ensure clearance from obstructions below the descent path.

At the start of the procedure, the controller will inform the pilot of the type of approach being provided, the runway in use, the angle of the nominal glidepath, and the procedure to be followed in the event of radio failure, unless that is published in the AIP. There is a Minimum Descent Altitude (MDA) which the pilot must not come below unless he can see either the runway or the approach lights, and he must calculate that from the obstacle clearance altitude or height which the controller will give him at the same time.

On a SRA, the controller will continue to provide vectors to keep the aircraft tracking towards the runway until the aircraft reaches a minimum range. That minimum range is given in the AIP, and is either 0.5 nm or 2 nm. He will also tell the pilot his range from the threshold. The pilot can work out the altitude he should be passing at each range, or usually the controller will tell him. The calculations are similar to those given in Chapter 8 on ILS, i.e. for a 3° approach path the aircraft height above touchdown should be the range multiplied by 300 ft.

When the aircraft has descended to the MDA, it is permissible for the pilot to fly the aircraft level until he reaches the published Missed Approach Point (MAP), looking for the required visual references. On reaching the MAP, if the references are not visible, he must start the missed approach procedure. In many cases, however, the MAP is designed to coincide with the descent path which the pilot would expect to follow, giving no opportunity for such level flight.

SRAs sometimes involve the aircraft approaching to a different runway to the one required for landing. This usually applies to light aircraft with low crosswind limit. Such an approach cannot be considered a 'runway approach' but is regarded as an 'airfield approach'. In that case, the aircraft must be flown level at the published 'circling minimum' once the airfield is in sight, to make a visual approach to land on the runway in use.

PSR can also be used to direct the pilot into a position from which he can make a pilot interpreted approach, such as an ILS. The director will guide the pilot in a similar fashion to a SRA until the aircraft is on a heading less than 45° from the final approach track and flying towards it.

PAR approaches

If a PAR is available, the surveillance radar controller (often called the 'director') hands over control to the final controller (often called 'talkdown') when the aircraft has appeared on the latter's PAR azimuth screen. This is usually just after a final vector to line the aircraft up with the runway centreline, and will be completed at least one mile before the aircraft reaches the glideslope.

The PAR aerials are positioned to one side of the runway, usually in a closed van which can be turned round to provide approaches to the opposite runway in about 30 min. The ICAO specifications for the system require it to be capable of identifying a $15\,m^2$ aircraft at a range of 9 nm, within a space bounded by a 20° azimuth sector up to 7° in elevation.

The display consists of the two screens, one above the other, represented in Figure 10.1, which the talkdown controller watches during the approach. He gives a continuous commentary which includes directions to put and keep the aircraft on the correct centreline and glidepath. Commentary words include 'above/on/below the glidepath' and 'left/on/right of the glidepath', as well as the ranges of the aircraft from touchdown.

Directions include 'turn left/right $x°$', as well as 'begin descent for a $x°$ glidepath' and 'take over visually'. The controller may also give instructions to 'reduce/increase rate of descent', but the pilot is expected to take corrective action on the basis of the glideslope information already provided. That glideslope information will normally include an actual distance above or below the glidepath, such as '50 ft below the glidepath'. During the approach, there should be no more than a 5 s gap between controller transmissions. The pilot is not normally expected to make any transmissions himself (although in the UK he may be requested to confirm the position of his undercarriage at a range of about 2–3 miles).

PARs are usually flown on QFE. This allows the pilot to compare his actual heights at the quoted ranges with those he expects, as a cross-check on the controller's instructions. The PAR procedure is shown at Figure 10.2. Before commencing the approach, the pilot calculates a decision height for

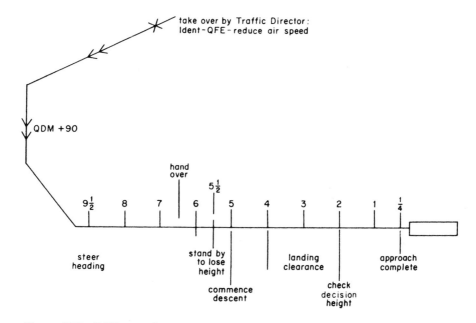

Figure 10.2 PAR procedure.

the procedure, and will be instructed to remind himself of it before he starts the descent on the glidepath. When he reaches that decision height, he must indeed make a decision, as he would at the equivalent point on an ILS approach. He decides whether to continue the approach or to climb away from the runway (go-around). He must commence a go-around and carry out the published missed approach procedure if he cannot see the required visual references.

Although the aircraft may have passed its decision height, the PAR controller will continue passing talkdown information, but not instructions, until either the aircraft is over the touchdown point or at some point before that. If the controller sees from his display that the pilot has gone around and is not continuing his approach, he will stop passing talkdown information and give the pilot instructions to help him with the missed approach procedure.

Monitoring ILS approaches

Both SRA and PAR can be used to monitor an ILS flown by a pilot. Like everything else in aviation safety, back-up is important, and pilots are recommended to take full advantage of all aids which can monitor the primary aid in use.

Advantages and disadvantages of PAR

Advantages

PAR requires no special equipment in the aircraft; the pilot only needs suitable flight instruments and a radio which can be tuned to the published PAR frequency. The director can provide safe and efficient separation between following aircraft. The PAR can be moved from one runway to its reciprocal easily, although it takes about half an hour.

Disadvantages

A dedicated controller is needed for the procedure. He can only talk one aircraft down at a time, so to permit full use of the procedure by several aircraft at once, separate controllers and communications frequencies are needed for each aircraft on final approach. This needs many frequencies, whereas ILS monitoring can be accomplished for several aircraft on one frequency. Because the wavelengths used are reflected by large rain drops, these weather returns require skill on the controllers part to identify the aircraft blip amongst the interference.

Sample questions

1. Which of the following transmissions may a pilot expect to make during a PAR final approach?

   a. 'Glidepath, descending'
   b. '3 greens'
   c. 'Approaching decision height'
   d. All of the above

2. Which of the following is not an advantage of PAR?

   a. It requires no special aircraft equipment
   b. It is a runway approach aid
   c. It allows a lower decision height than an SRA
   d. When the runway is changed, the PAR can be used on the new runway instantly

3. A controller may ask a pilot to carry out one or more actions when the pilot requests a radar service. Which of the following would not be requested?

   a. A series of turns
   b. A VOR/DME position
   c. 'Squawk ident'
   d. All the above may be requested

4. What is a radar vector? Choose the most accurate answer.

   a. The heading the controller wants the pilot to steer
   b. The safe track down which the aircraft should approach the runway
   c. The line on display screen which represents the timebase
   d. The 'blip' on the radar screen

5. What do we call the point at which the pilot must initiate a 'go-around' from a Surveillance Radar Approach?

   a. Decision height
   b. Reference point
   c. Minimum descent altitude
   d. The missed approach point

6. ICAO states that a PAR must be able to identify an aircraft with a reflecting area of $15\,m^2$ over certain sectors in azimuth and elevation. What are these sectors?

a. 20° in azimuth, 7° in elevation, out to 18 nm.
b. 20° in azimuth, 7° in elevation, out to 9 nm
c. 40° in azimuth, 7° in elevation, out to 18 nm
d. 18° in azimuth, 9° in elevation, out to 20 nm

7. Which of the following would have the shortest required maximum range?

   a. TAR
   b. ASDE
   c. PAR
   d. RSR

8. The picture below represents a PAR display with an aircraft 'blip'. What information would the pilot expect from the controller with regard to the aircraft's position?

   a. Left of centreline, above the glidepath
   b. Right of centreline, above the glidepath
   c. Left of centreline, below the glidepath
   d. Right of centreline, below the glidepath

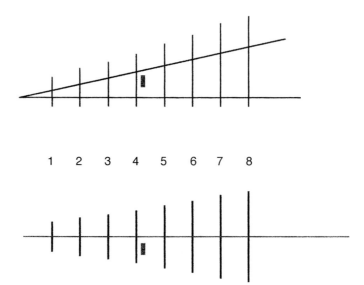

Chapter 11
Displays

Introduction

Modern cockpit environments in commercial aircraft have removed the traditional layout of analogue instruments showing individual items of information. Instead, the pilot is presented with one or more electronic screens which provide the information he needs. These screens can usually perform more than one function, either simultaneously or individually as required by means of switches. Most navigation aids have their information displayed on some form of electronic screen in modern airliners, and this chapter will explain some of the details of the common types of these multifunction display.

Although there are various industry acronyms used to describe the displays, they can basically be divided into two types. As mentioned in Chapter 9, the cathode-ray tube (CRT) was originally taken from laboratory experimentation into cockpits to show the reflected signal returns in radar systems, and progressed to the full colour screens common in desktop computer monitors, and familiar to most people today. The other basic type is the liquid crystal display (LCD), which started as a simple method of showing figures in a tiny window and progressed again to full colour in the 'flat screens' used in laptop computers. The last part of this chapter is concerned with head-up displays.

Cathode-ray tubes

Basic principle

The screen itself is one end of a sealed glass tube, coated on the inside with substances (phosphors) which glow when electrons strike them. The other end of the tube has a substance such as barium oxide, with electrical connections from the control circuits. When a current flows from the control circuits, the oxide is heated. The heated oxide releases electrons from this negatively charged 'cathode', or 'electron gun'; they flow to the screen at the other end which has a positive electrical charge. When the electrons strike the fluorescent screen, the phosphors emit light (glow), and continue to glow for a finite time depending on the particular phosphor.

The number of electrons released at the cathode (which is heated to excite the electrons and give them more energy) depends on the voltage in the control circuit. In order to focus the electrons on to the screen, they pass through a negatively charged tube, called a 'grid'. This 'repels' the electrons into a thin stream. More positive and negative charged plates and tubes are fitted to refine the focus, much as glass lenses focus visible light beams.

The electron stream is designed to make a spot on the screen. It passes across the face of two pairs of plates on its way. These plate pairs are set up at 90° to each other, and by inducing an electrical field between the pairs of plates, the electron stream can be attracted (or repelled) across the screen towards (or away from) the plate. The vertically oriented plate, which moves the stream horizontally across the screen, is referred to as the 'X' plate, and the horizontally oriented plate, which moves the stream vertically across the screen, is referred to as the 'Y' plate. (Mathematicians refer to the horizontal axis of a graph as the 'x' axis, and the vertical axis as the 'y' axis). The deflection, and some of the focusing, is usually carried out more efficiently by magnetic fields from electromagnets outside the tube. A pictorial representation of a CRT is shown in Figure 11.1 below.

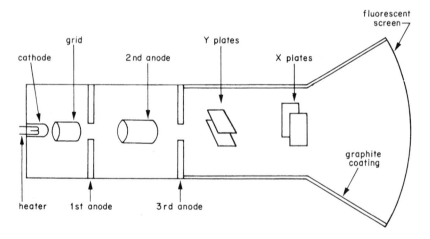

Figure 11.1   CRT components.

Radar scopes

Varying the voltage across the 'X' plates produces a horizontal line on the screen. As mentioned in Chapter 9, on early radar 'scopes', this line formed a 'timebase' moving across the screen to reach the far side at a time equalling the prp of the radar pulse, proportional to the maximum range of the radar.

The reflected signal excited the 'Y' plates when it was received, producing a 'blip' on the screen at a distance proportional to the range. A simple scope is shown in Figure 9.2. PPIs involve rotating the 'X' and 'Y' plates with the antenna.

To provide topographical reference for controllers on PPI displays, the basic radar picture is often projected on to a further screen which has a map of the area permanently marked on it. This 'video map' is the normal display in control towers and area control centres. The map on modern displays is generated by computer.

Raster screens

Cockpit display CRTs are similar to television screens. In those, the timebase is swept across the screen in lines. At the end of the line, the beam is switched off while it is brought back to the start of the next line. The picture is drawn from top to bottom, as shown in Figure 11.2. The control signals alter the brightness of each part of the line to produce the total picture. Instead of scanning from top to bottom line by line, some screens scan in an 'interlaced' fashion. This involves scanning every second line, then when the scan reaches the bottom of the screen it goes back and scans the missing lines.

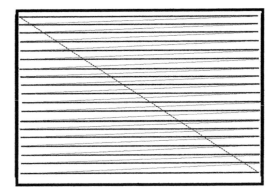

Figure 11.2   Raster scan.

Colour CRTs

Colour displays frequently involve three different types of phosphor to produce three different colours on the screen. These mixed in various ratios produce all the colours visible to the human eye. Each primary colour has is own cathode, and the different phosphors are painted in small dots on the screen. In front of the screen is a 'shadow-mask' with many small holes. The pattern of dots and shadow holes is such that each colour's electron gun can only strike its own phosphor when it is activated.

It is possible to produce the same effect in a different fashion, by using slots instead of holes in the shadow-mask. Other forms of colour CRTs use a single electron gun and three layers of phosphors on the screen. The different colours are produced by controlling the acceleration of the electron beam before it strikes the screen. The faster the electrons are moving, the deeper into the phosphors the beam travels, producing the required colours.

Fibre optics

Rather than directing the electron beam through a shadow-mask on to a phosphor screen, some systems use coloured optical fibres to transmit the light.

Liquid crystal displays

The cathode-ray tube uses a lot of space, and is relatively heavy. When fitted to an instrument panel, its depth can cause problems. The screen itself is usually gently curved to reduce distortion, and that can be considered to intrude into the cockpit. To overcome these disadvantages, liquid crystal displays, 'flat screens' are becoming more and more common in aircraft cockpits. These LCDs also reduce the avionics heat loading.

Light is emitted at the back of the screen, and passes through one or more polarising filters (see Figure 11.3). The polarised light passes through a layer of liquid crystal cells, each of which can accept an electrical charge from a computer graphics adaptor. When charged, the crystals change shape into spirals by an amount proportional to the charge. The polarisation of the light leaving the crystal is therefore changed by that amount.

A further polarising filter, at 90° to the first, then prevents light which has not been twisted from continuing to the front panel of the screen. Fully

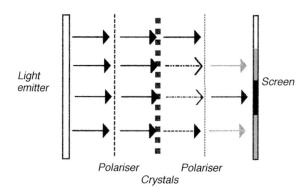

Figure 11.3   LCD process.

charged and twisted cells allow their light to reach the screen at full intensity. Only a small proportion of light which has been slightly twisted will reach the screen. Patterns of light at varying brightness can then provide a single colour image which can display a 'black and white' image.

Colour LCDs

To produce colour on the screen, the light from each cell passes through one of three colour filters between the two polarising filters. When coloured light from each cell is seen mixed with other coloured light from neighbouring cells, it appears as a combination colour. An instrument display on a LCD screen is shown in Figure 11.4.

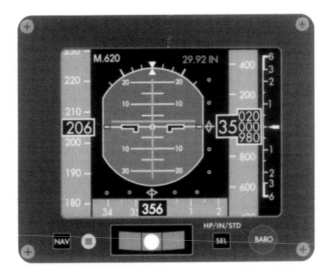

Figure 11.4 EADI display (courtesy Meggitt Avionics).

Head-up displays

The previous screen types are placed in the instrument panels, and are used as displays for many of the equipments described in the following chapters, as well as for the Electronic Flight Instrument System (EFIS). However, military aircraft have for some time used Head-up Displays (HUDs) which allow the pilot to see representations of the flight instruments while looking out of the windscreen.

The advantages of such an arrangement became obvious to civilian pilots making approaches to land in poor visibility. They are able to follow ILS or other instrument directions while looking ahead for the runway. At the time of writing, suggestions have been made that head-up displays should allow aircraft to make category III approaches from equipment which at present is limited to category II.

The HUD is basically a transparent glass plate placed in the pilot's line of vision on which the symbology of the required instrument is projected from behind and above the pilot's head. The projection is focused at infinity, so that the pilot's eyes do not have to refocus from the view outside to the display. To use the HUD properly, the pilot's eyes must be aligned carefully with the screen. This makes the seating position of paramount importance. The screen itself needs to be coated with a substance to allow the projected image to show clearly, while still giving the pilot clear vision through it. Figure 11.5 shows a typical system, consisting of screen and projector, mounted in the cockpit structure. Figure 11.6 shows some of the symbols seen on a HUD screen.

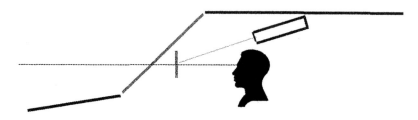

Figure 11.5   HUD components.

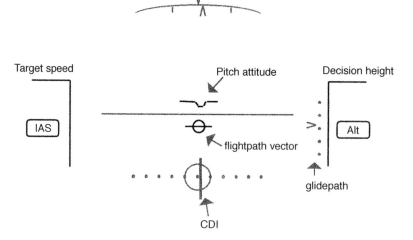

Figure 11.6   HUD screen.

Chapter 12
Airborne Weather Radar

Introduction

We have already looked at the basic radar theory. The most common 'primary' radar system used in aviation is the airborne weather radar (AWR) fitted in many commercial aircraft. It is also used for navigation. A sector scanning antenna is fitted, usually in the nose of an aircraft, behind a cover called a 'radome' made of a material which is transparent to radar frequencies. The scan is horizontal around the aircraft's nose providing information relative to its heading. The shape of the beam can be varied to provide the particular information required by the crew.

Principle of operation

Weather radars were originally introduced because of the dangers to aircraft flying in the troposphere from cumulonimbus clouds, as listed in *Ground Studies for Pilots: Meteorology*. Most of the features of cumulonimbus clouds are difficult to identify at safe ranges. However, the strong upcurrents in such clouds can support large raindrops, so radars with wavelengths which could be reflected by these large raindrops, but not by smaller ones, were developed.

As discussed in Chapter 9, a frequency of approximately 10 GHz, giving a wavelength of about 3 cm, is used in order to provide adequate reflections from these large raindrops. Hailstones, one type of hazard from cumulonimbus clouds, are also reflected satisfactorily.

Since the system is designed as a warning device, accurate range and angular discrimination is not needed. It is therefore possible to transmit long pulses, which provide a high peak signal strength for a low average power consumption. The antenna need not be large, so a small parabolic aerial which can fit in the nose of even a light aircraft can provide a satisfactory beam width of about 3°.

The conical beam

The narrow conical beam (traditionally between 3° and 5° beam width) can be tilted up or down as required by up to 15°. By measuring the vertical angles at which the reflection is received, a calculation can be made of the relative height of the rain-bearing cloud to the aircraft. If appropriate, the aircraft altitude can be adjusted to avoid the worst of the weather, although it must be emphasised that a lack of reflected signal does not necessarily mean a lack of hazard! The antenna is normally stabilised gyroscopically by connection to the aircraft's attitude reference system, so that the angles measured are relative to the horizontal.

The intensity of the reflected signal (the 'return'), can be indicated on the display. By the shape and intensity of that return, a pilot can judge the likely severity of the weather ahead. If the reflecting drops or hailstones are relatively far apart, much of the signal will travel between them, and the return will be relatively weak. If however, the reflecting drops are close together, which would be the case in a severe cumulonimbus cloud, most of the signal will be reflected from the nearest point to the antenna, and the return will be strong. Of course, once the signal has been reflected, it cannot continue beyond the point of reflection, so any object at greater range than such a strong return will be hidden in this 'shadow area'.

Not only raindrops and hailstones will reflect the signal. Any object of a similar or larger size will do so, and it was found that details and shapes could be identified. Possible conflicting air traffic could be seen, but could be confused with weather returns. However, when the beam touched the ground, reflections from coastlines were identifiable, so it was a short step to using the weather radar for navigation purposes by comparing the display with a map.

The 'cosecant squared' beam

A narrow beam is not ideal for mapping purposes. It may be compared to using a torch with an equivalent narrow beam to light the way ahead in total darkness. Especially when looking relatively close to the aircraft, the small area illuminated gives insufficient information. To cover a larger area, a different vertical shape for the beam was devised. This can be selected by a switch in the cockpit to change either the shape of the antenna or the position of the feeding waveguide. The resultant fan-shaped beam allows reflections from objects at a greater spread of distances to be received from the antenna.

However, close objects give much stronger returns than distant ones, so the signal pattern has to be altered to provide a fairly equal return from objects at different ranges. This is achieved by varying the strength of the signal in relation to the square of the cosecant of the angle of depression of each part of the fan-shaped beam. Hence the close range mapping beam has become known as the 'cosecant squared' beam.

Weather displays

Black and white

Normal cathode-ray tubes provide a single colour (or 'black and white') display of the returns. The pilot is not normally interested in weather elsewhere than along his intended flight path, so traditionally the screen covers an angle of about 75° either side of the aircraft's heading, although several extend further. White returns indicate the reflecting raindrops or hailstones. Modern black and white displays can be reversed to show the returns as black on white, which in certain lighting conditions makes the shape of the return easier to identify.

Large concentrations of raindrops or hailstones will cause most of the signal to be reflected. Although weak echoes will appear faintly, a simple display has no way of distinguishing particularly strong returns, which would indicate the presence of a cumulonimbus cloud, from the moderately intense returns from ordinary, non-hazardous, rain-bearing clouds.

A system has been devised to show levels of intensity by processing the returns as they are received. Any part of a return above a certain strength level is phase changed. This means that a return entering the antenna at more than the datum strength will be processed, and the part above the selected level being at negative phase added to the positive phase of the part below reduces the displayed return. A return at twice the datum strength, representing a large concentration of water droplets, actually appears to have no strength at all, and no paint is made. The processing of a very strong return produces more negative phase than positive phase signal, and the screen does not display this negative phase signal. However, it is surrounded by the weaker returns, so very intense signal returns appear as a black centre with a white surround (or vice versa if the display has been reversed). This display is called an 'iso-echo' or contour display, and usually has to be selected by a switch on the control panel. A reversed image iso-echo display is represented in Figure 12.1 below.

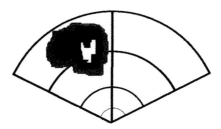

Figure 12.1 Iso-echo (reversed) display.

Colour

More modern displays use different colours for different return intensities. Relative return strengths can be shown either on a dedicated instrument face or as part of a larger horizontal situation display, for example an EFIS screen. The colours are traditionally the same as those used on television rainfall radar pictures – green for a low intensity return, yellow for a moderate return, and red (or magenta) for a very strong return. This sequence can be considered as being similar to that of traffic lights; red for 'danger', amber for 'warning', and green for 'no hazard at present but possibly changing'.

Ranges

The range of a return can be measured against range rings on the display. These can be varied by the pilot on his control panel, as can the scale of the displayed picture itself. Traditionally, a 120–150 mile range (300 on EFIS displays) is used for initial detection of the weather, but a larger-scale picture is preferred for calculations. In addition to the range rings, a ring at a range equivalent to the aircraft's height is also displayed.

Filters are applied to the received returns. Those from within about 25 nm are displayed normally, but as the time from the transmission increases, the sensitivity of the receiver is proportionately increased. This allows iso-echo or colour discrimination at longer ranges, up to about 70 nm, and is called 'swept gain'.

Turbulence indication

As stated in the introduction, AWRs do not show the hazards contained in a cumulonimbus cloud, but only the reflecting precipitation. Pilots have to learn to interpret the display to identify the areas where the hazards are most likely. Commercial airliners usually regard turbulence as the most undesirable effect of a cumulonimbus cloud. The most severe turbulence is commonly found in the area where the intensity of the precipitation (and therefore the return intensity) changes most rapidly. The 'intensity gradient' is strongest where the white part of an iso-echo return is narrowest, or where the yellow or green parts of a colour return are narrowest in a colour display.

The shape of the return has also proved to be an indication of turbulence, and on older displays may be the only one. Tornadoes (see *Ground Studies for Pilots: Meteorology*) often produce 'hooking' shapes on a screen, where the precipitation is being swirled around by the air movement (see Figure 12.2). Such 'hook' shapes should be avoided, as should any returns with sharp edges, or those changing their shape rapidly.

Figure 12.2 'Hook' echo on a reversed display.

Avoidance recommendations

In an Aeronautical Information Circular (AIC), the CAA advises pilots to avoid such returns by varying distances. Below 20 000 ft, pilots should avoid returns with sharp edges or strong intensities, or those showing a close gradient change on an iso-echo or colour display, by 5 nm. Pilots should avoid echoes showing protrusions such as hooks, or those rapidly changing their shape, height or intensity, by at least 10 nm. Above 20 000 ft, the AIC recommends avoiding all echoes by minimum distances depending on the aircraft's altitude. From 20 000 to 25 000 ft, avoid all echoes by 10 nm. From 25 000 to 30 000 ft, avoid all echoes by 15 nm. Twenty miles is the figure when flying at a pressure altitude above 30 000 ft.

Relative height calculation

Once a return has been detected, the beam can be moved up and down by up to 15° to determine the relative height of the worst of the weather, and a possible vertical avoidance manoeuvre. The 1 in 60 rule, as detailed in the volume on *Navigation* this series, is a simple method of making the calculations. If the beam just misses the precipitation when it is angled upwards at 1° at a range of 60 nm, one could calculate that the top of the precipitation is 1 mile (6000 ft) above the aircraft.

However, if the beam has a vertical width of 3°, the bottom of the beam is 1.5° below the centre. In this example, the top of the precipitation is actually 0.5°, and therefore 0.5 miles or 3000 ft below the aircraft. This can be seen in Figure 12.3 below. Again, the CAA advises in its AIC that returns should be

Figure 12.3 Weather radar beam width.

avoided vertically by a minimum of 5000 ft. It is not a good idea to fly below a cumulonimbus return.

Mapping display

At long range (over about 50 miles) the conical beam gives adequate mapping information when it is directed slightly downwards. Weather information can also be obtained at the same time. At shorter ranges, the operator must select the cosecant squared beam. As on the weather display, there is usually a 'height ring' shown at a range equal to the aircraft's height above the ground, as well as rings indicating the slant ranges. These are displayed as a proportion of the maximum range selected by the pilot.

Coastlines are the features most commonly used for navigation, because the radar beam is reflected away from flat water features, and little energy comes back to the receiver. Land surfaces, however, are irregular and many small objects reflect the radar wave back. This produces a pattern of white return from land, and black 'emptiness' from water, which makes patterns on the screen similar to the map shape of coastlines or lakes. Over large land areas, the difference in reflectivity between various surfaces can be used to identify features. Hills, buildings, rough terrain and groups of metal objects provide good reflections and powerful returns, but hill shadows can be confused with water features by their darkness. Doppler radar techniques may be used in some equipments to provide finer resolution of ground features.

The returns from ground features vary with the angle, horizontal or vertical, at which the beam strikes them. They also change with weather and season, and often bear little obvious relation to features on a normal map. Experience and training are required to use airborne radar for navigation over land.

The mapping radar may be used for navigation, and in examinations. In these cases, the display will show a prominent object at a certain range and angle from the centre of the display. The important facts to remember are that the range shown will be slant range, and that the radar is aligned with the aircraft's longitudinal axis. The direction of the centre of the display will be the aircraft's heading, not its track. To plot an aircraft's position a navigator must first find the aircraft's true heading, then plot the reciprocal, allowing for convergence if necessary. In the example in Figure 12.4 below, if the maximum range selected is 75 nm, a prominent headland is situated at 25° left of the aircraft's heading at a range of 50 nm. If the heading is 090° true, the headland bears 065° true from the aircraft, so the aircraft can be plotted at a bearing of 245° true from the headland, at a range of 50 nm.

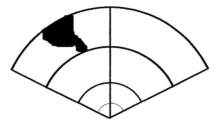

Figure 12.4   Map display (reversed image).

Cockpit controls

A control panel of a CDU is depicted in Figure 12.5. The brightness (BRT) knob may be replaced by a 'colour intensity' control. There is a rotary control to select the maximum range of the display, and the range rings will appear on the display at proportions of that maximum range. The gain control is usually left at 'automatic' in the WEAther mode, but when using the radar for mapping the operator should adjust the gain manually by use of the knob to obtain the best possible picture. The function control on this unit has positions to display weather (WX), ground returns (MAP) and a combination of weather and TCAS information (WX/T), as well as a test facility. Additional functions may include the display of windshear warnings, as described below.

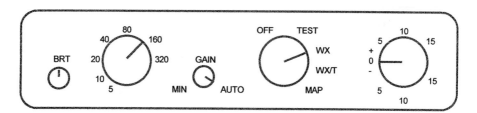

Figure 12.5   AWR control unit.

The tilt control on the right should be used to search for the most likely and most severe weather returns. These would usually appear in the middle of a cumulonimbus cloud, where the updraughts are strongest, so if the aircraft is at high altitude, the most intense returns would appear below the aircraft, and above it if flying at low altitude. The crew can tilt the antenna accordingly. Another reason for adjusting the tilt would be to remove unwanted ground returns from the weather display.

Requirements

JAR-OPS 1.670 requires that certain aircraft must carry airborne weather radar equipment 'when operating at night or in instrument meteorological conditions in areas where thunderstorms or other potentially hazardous weather conditions, regarded as detectable with airborne weather radar, may be expected to exist along the route.' These aircraft are:

- All pressurised aeroplanes.
- Unpressurised aeroplanes with a maximum certificated take-off mass of over 5700 kg.
- Unpressurised aeroplanes with a maximum approved passenger seating configuration of more than 9 seats.

Unpressurised propeller-driven aeroplanes lighter than the above and with fewer seats may be operated with other equipment capable of detecting thunderstorms, for example the lightning detector as described below, if the equipment is approved by the relevant authority.

Windshear detection

Turbulence associated with windshear is a major safety problem during the approach and departure phases of flight. Even at high altitudes, windshear in frontal zones (see *Meteorology* in this series) can have a considerable affect on passenger safety. Systems have been developed which compare airspeed information from the Air Data Computer (ADC) with groundspeed information from the navigation computer. This comparison can detect changes in wind affecting the aircraft. Pressure altitude information from the ADC can provide information about rates of climb or descent, and when coupled to airspeed (and air temperature) changes can give an indication of the effect that a gust is having on the aircraft.

As described in Chapter 21 of this volume, some TAWS computers will alert the crew if an increasing headwind (or decreasing tailwind) and/or a severe updraught exceed a defined threshold. These are characteristics of conditions which might be expected just before an encounter with a microburst (again see *Meteorology*). A decreasing headwind (or increasing tailwind) and/or a severe downdraught are characteristic of conditions that might be experienced within a microburst itself, or just afterwards, and result in a more urgent warning.

The alert and warning thresholds depend on available climb performance, flight path angle, airspeed changes, and fluctuations in static air temperature. A windshear alert on a weather radar display will often take the form of bands of colour across the screen in the aircraft's direction of flight.

Lightning detectors

For light aeroplanes, alternative systems have been developed, some of which have been incorporated in systems for larger aircraft also. One commercially available system uses a small external aerial incorporating directional antennas (on a similar principle to ADF) to detect electrical discharges (lightning) from thunderstorms in line-of-sight range. The system computer uses the signal strength of this discharge to determine not only the direction from which it has come, but also its range. The equipment can then display each discharge on a screen.

By keeping each discharge in the computer's memory, a map can be built up of the discharges over a period of time. This shows the areas of most frequent discharge, which can be assumed to be the most intense thunder-storms. The thunderstorm map may be shown on an independent screen or as part of any other horizontal situation display. In some installations, the pilot may have the choice of displaying either the normal map of historical information, or a screen which shows only the discharges which have occurred from the moment he started a 'stopwatch' on the equipment. This facility is useful if the pilot is committed to flight through a stormy air mass and wishes to know where the storm activity is reducing. An area of reducing activity is likely to be less dangerous than one where the activity is increasing. Figure 12.6 shows such a display of 'strikes' on a simple EHSI MAP display fed from a navigation computer.

Figure 12.6   Discharges on nav display (courtesy BFGoodrich Aerospace).

Summary

Principle	Large raindrops reflect 3 cm radar beam
Frequency	SHF, 9–10 GHz
Accuracy	Azimuth resolution and 'stretching' depends on beam width
	Range resolution and 'stretching' depends on pulse width
WEAther	Conical beam (approximately 3° wide) moving in azimuth, tiltable ± 15°
Range	120–150 nm mono, up to 300 nm EFIS
	Swept gain gives equal intensity over first 25 miles, reduces thereafter
ISOechos	Echoes over a limit are inverted (monochrome) or differently coloured
	Inverted means 'Black Hole' for strong echo
	Coloured means RED strongest, YELLOW medium, GREEN normal
Turbulence	Worst where colour (contour) changes fastest
Storms	Avoid strange shapes and sharp edges
	At high levels avoid all echoes by 20 nm
Cloud height	Use 1 in 60, remember beam width
MAP	Conical for long range
	Broad fan for close range (cosecant squared beam = power increase as distance up to 70 nm)
Gain	Auto for normal weather
	Manual gain in MAP
Reflections	Hill, buildings, metal, rough terrain = excellent reflection
	Sea, hill shadow = dark

Sample questions

1. On an EFIS AWR picture, what colour depicts the strongest echoes?

    a. Green
    b. Yellow
    c. Blue
    d. Magenta

2. What is the 'cosecant squared beam' of an AWR used for?

    a. To measure cloud height accurately

b. To avoid strong close range ground returns swamping the MAP picture
c. To give more detail at ranges over 60 nm
d. To remove weather returns from the MAP picture

3. Azimuth resolution on a AWR depends on:

   a. Pulse width of the transmission?
   b. Frequency of the transmission?
   c. Beam width of the transmission?
   d. PRF of the transmission?

4. Why is a wavelength of 3 cm used for AWR?

   a. To allow the antenna to fit in the nose of the aircraft
   b. To receive reflections from large water drops
   c. To receive reflections from small as well as large clouds
   d. To keep the weight of the airborne equipment to a practical level

5. Which of the following contour indications on AWR would most likely indicate turbulence?

   a. A large red area
   b. A large yellow area
   c. A large green area
   d. Colours changing very rapidly with distance

6. What is the wavelength of a typical airborne weather radar?

   a. 10 mm
   b. 30 mm
   c. 10 cm
   d. 30 cm

7. What is the normal maximum range of an AWR displayed on EFIS?

   a. 20 nm
   b. 120 nm
   c. 200 nm
   d. 300 nm

8. An AWR echo at 30 nm when the beam is level with the aircraft is lost when the centre of the beam is tilted upwards at an angle of 2°. How high above the aircraft is the top of the rain?

   a. 600 ft

    b. 1500 ft

    c. 3000 ft

    d. 6000 ft

9. Looking at AWR returns on an EFIS screen, where is turbulence most likely?

    a. Where the colour changes rapidly

    b. Where hook-shaped echoes are seen

    c. Both (a) and (b)

    d. Turbulence cannot be detected

10. Why is the mapping beam of AWR sometimes referred to as the 'cosecant squared beam'?

    a. Time of scan depends on the cosecant^2 of the angle from the aircraft's nose

    b. Range is found by measuring the cosecant^2 of the angle of the beam

    c. The cosecant^2 of the angle of the beam is proportional to the aircraft's speed

    d. Transmitter power is proportional to cosecant^2 of the angle from horizontal

11. By how much should a long range jet pilot avoid strong echoes on his AWR when flying at cruise altitude?

    a. By a minimum of 2 nm

    b. By a minimum of 5 nm

    c. By a minimum of 10 nm

    d. By a minimum of 20 nm

12. What does 'swept gain' mean in an AWR?

    a. Transmission power increases as range increases

    b. The returns are stronger at large angles from the aircraft heading

    c. Transmission power increases as the cosecant^2 of the angle from horizontal

    d. The receiver increases sensitivity with time

13. What is an 'iso-echo'?

    a. A black indication in the middle of a white radar return

    b. A strong echo showing turbulence

    c. A green indication on an EFIS display

    d. A strange shaped return on the AWR screen

14. Flying at FL 270, an AWR weather return at range 40 nm is identifiable when the centre of the beam is tilted between $\pm 1°$ and $-6°$. What is the height of the base and top of the rain-bearing cloud?

    a. Base FL 30, top FL 210
    b. Base FL 30, top FL 290
    c. Base FL 90, top FL 210
    d. Base FL 90, top FL 290

15. Given the AWR mapping indications shown below, the aircraft heading 270°(T) at 30 000 ft and the range scale at 150 nm, what is the correct interpretation?

    a. A headland bears 300°(T) at a range of 300 nm
    b. A headland bears 030°(T) at a range of 300 nm
    c. A headland bears 300°(T) at a range of 100 nm
    d. A headland bears 030°(T) at a range of 100 nm

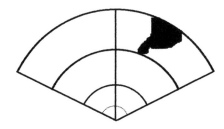

Chapter 13
Doppler Radar

The Doppler principle

Often in schools the 'Doppler' principle of sound waves is described by considering the noise of a train whistle as the train passes an observer. While the train is approaching, the frequency of the whistle appears to be higher than it really is, and as the train is moving away from the observer, the frequency appears lower. Radio waves can be said to behave in a similar fashion.

A signal from a radio broadcast transmitter moves away from the transmitter in all directions at the same constant speed (3×10^8 m/s). The radio wave itself leaves the aerial in the same form in all directions. However, if we consider a portion of the wave form generated from a moving transmitter, for example the 2.25 cycles in Figure 13.1 below, by the time the last part of the portion leaves the aerial, the aerial itself will have moved relative to the first part. The portion of the wave will appear to have shortened in the direction of the aerial movement. A stationary receiver in the direction of travel, as in Figure 13.1, will 'hear' more cycles per second, therefore a higher

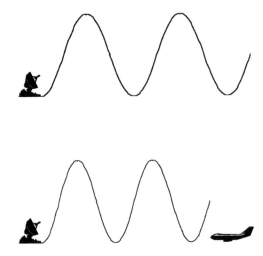

Figure 13.1 The Doppler effect.

frequency than that transmitted. By the same analogy, it will appear to have lengthened in the opposite direction. The same would apply if the receiver were moving and the transmitter was not, or if both were moving. The received frequency will change in relation to the relative speed between the transmitter and the receiver.

For a normal pulse radar, this may be considered a disadvantage, in that a moving target will reflect a frequency slightly different from that transmitted. It means the radar receiver must have a fairly broad bandwidth to accept a variety of reflected frequencies around the original. However, the principle can be used in other applications to expand the value of electromagnetic waves in navigation.

One benefit of a Doppler radar lies in its ability to transmit continuous wave (CW) signals rather than pulsed. Effective power is increased because the pulse width is not limited. In CW signals, the transmitted frequency is blanked out at the receiver, and only signals reflected from those targets which have relative movement towards or away from the radar unit, and which are returned at a different frequency, are processed.

Moving targets

A fixed Doppler radar can detect any target which is moving towards or away from it by transmitting a constant frequency and comparing the frequency of the received, reflected signal. While it is transmitting, it cannot receive exactly the same frequency, and so any target which is neither approaching nor receding will not be detected. However, that means there is no 'ground clutter' or weather return to interfere with the picture. If the transmission is constant (CW), less power is needed for detection at the same range. Of course, a Doppler radar cannot indicate the range of the target, but if used in conjunction with a primary pulse radar, which can, it can filter out the signals which show no 'Doppler shift'. As a result, a display fed from both systems, and indicating only those targets which respond on both, will only show moving targets. The system has a 'moving target indicator' or MTI.

The problem with a MTI is that a target which is moving at slow speed, such as a balloon, will be filtered out. Any target which is flying tangential to the receiver (at 90° to the track to the antenna) will also have no Doppler shift, and so will not be visible either. However, the system was used for some time for Air Traffic Control with fair success.

Relative speed measurement

The actual 'Doppler shift' in frequency from transmission to reception of the wave, can be used in a formula to discover the relative speed between the transmitter and the receiver. The basic formula is:

$$f_{\mathrm{d}} = \frac{V}{\lambda} \quad \text{or} \quad V = \lambda f_{\mathrm{d}}$$

where f_{d} is the change in frequency due to the Doppler effect (i.e. receiver freq – transmitter freq), V is the relative velocity between the transmitter and receiver and λ is the wavelength of the original signal. The formula can also be written $f_{\mathrm{d}} = VF/c$ where F is the signal frequency.

However, when the effect is found in a radar, the target is moving relative to both the transmitter and receiver, so the received frequency will be affected double by the Doppler effect. The formula for a radar signal then becomes:

$$f_{\mathrm{d}} = \frac{2V}{\lambda} \quad \text{or} \quad V = \frac{\lambda f_{\mathrm{d}}}{2}$$

Airborne Doppler

The Doppler principle has been available for a considerable time in an airborne (self-contained) navigation system, with the advantage that it does not depend on the serviceability or availability of external navigation aids. It may still be used to provide information to a navigation computer. In principle, the airborne Doppler equipment sends a continuous radar beam forward of the aircraft, measures the Doppler shift (frequency change) of the received signal, and computes the aircraft's groundspeed from the Doppler formula. In practice, four (or often three) beams are used, pointed at an angle downwards to reach the ground, and from the received signal frequencies the computer can calculate drift as well as groundspeed.

For accuracy, the system should produce the maximum frequency change, so a beam should be as near horizontal as possible. However, at a small depression angle, much of the beam is reflected away from the receiver, as in Figure 13.2 below, so a larger depression angle is required, usually about 60°.

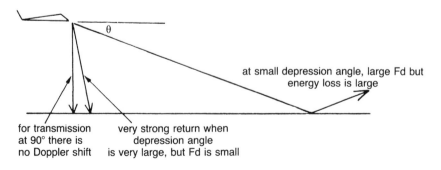

Figure 13.2 Depression angle.

The beams typically have a width of $1°$ to $5°$. The actual Doppler shift will be less than from a direct reflection, in fact it will also depend on the cosine of the depression angle θ, giving a reflected Doppler shift for one beam (f_d) of $2VF \cos \theta / c$.

It is unusual for an aircraft to fly at a constant attitude for long, so the depression angle would continually change. To overcome such variations, a beam pointed backwards at the same depression is used, and that shift is also measured. The frequency received by the rear receiver would be less than that from the transmitter, whereas that from the front would be more, although the actual numerical change would be the same if it were not for the variation in depression angle. If the aircraft pitches nose up, the effective depression angle of the front beam will decrease, and that of the rear beam increase, as seen in Figure 13.3.

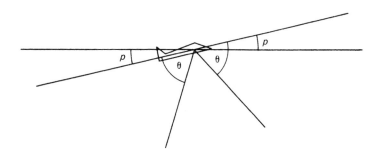

Figure 13.3   Effect of pitch.

Having measured both shifts, a positive shift from the front and a negative shift from the rear, the computer subtracts the rear shift from the front shift, producing a large positive shift. When that total shift is fed into the formula and divided by two, it is as if the average depression angle were almost constant. This also compensates for minor frequency errors in the trans-mitter, and in addition takes account of rising or falling ground beneath the aircraft. This double beam system became known as a 'Janus array'.

An aircraft experiences drift, and moves along a track which is different from its heading. An error would occur in its calculated groundspeed, as the equipment would only measure the speed along that heading. This can be prevented by fitting two beams pointed at an angle out from the heading, rather than one. The two shifts can be added together and divided by two in order to produce an average shift along track. Again the formula has to be altered, not only to divide the total shift from the forward beams by two, but also to compensate for the angle out from the heading. In simple practice, four beams are used as in Figure 13.4, two angled forward and two angled rearwards, although only three are strictly necessary because the computer

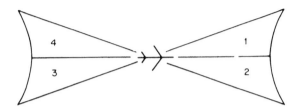

Figure 13.4   A four-beamed pattern.

can make the calculations needed. The horizontal angle (φ) is typically about 10°.

The antenna system usually sends out its beams in a sequence of pairs, front right and rear left together, then front left and rear right. The computer can determine the difference in shift, therefore speed, left and right of the heading, and by combining that with the groundspeed a drift angle can be calculated. In fact, most systems use the difference in shift to turn the aerial itself until there is no difference. The physical angle between the antenna system and the aircraft is the drift angle.

The actual groundspeed can be calculated from the rather complex formula

$$f_\text{d} = \frac{4VF \cos \theta \cos \varphi}{c}$$

$$\text{or} \quad V = \frac{cf_\text{d}}{4F \cos \theta \cos \varphi} \quad \text{or} \quad V = \frac{f_\text{d}}{4\lambda \cos \theta \cos \varphi}$$

Characteristics of airborne doppler

The accuracy and reliability of the Doppler system depends on the quality of the reflected signals. That in turn depends on the surface which is reflecting the signal. A smooth surface, such as calm water, will reflect poorly, because most of the energy will be reflected away from the receiver. Vertical land features, or a rough surface such as sand with a grain size close to the wavelength of the signal, will reflect well. Rolling pasture or swelling water will usually give a good signal.

Frequency

There are two frequencies in use by commercially available systems. Both of these, 8.8 GHz and 13.5 GHz, are in the SHF band to allow a narrow beam width and accurate calculations using the formula.

Accuracy

The four-beam (Janus) system is regarded as having an accuracy of 0.1% of the true groundspeed, and 0.1° of drift. The ICAO standard for a Doppler system is to demonstrate an accuracy of 0.5% of both indicated groundspeed and indicated drift.

Errors

- If the aerial is slightly misaligned with the aircraft axis, errors will occur.
- If the reflecting surface, the sea, is itself moving, the relative speed will not be true groundspeed. This is often called 'sea movement' error.
- If the aircraft is climbing or descending, the system will detect movement, although the Janus array minimises this.
- There is an error called 'sea bias'. On a flat surface, the closest part of the beam will reflect more strongly than the farthest part, because it has a greater depression angle. The calculated groundspeed will then be less than true groundspeed, because the formula assumes the depression angle of the centre of the beam. A land/sea switch is fitted, which when operated, increases the indicated groundspeed by 1%.
- As an extension of sea bias, if the water is flat calm, not enough energy will return. A memory circuit maintains the movement of the indications at the previous rates (deduced reckoning or DR).
- The drift can be measured accurately, but if the accurate drift is applied to an inaccurate compass heading, the inaccuracy will be carried over to the calculated track.

Doppler derived position

The raw Doppler information would normally be fed to a navigation computer to calculate the aircraft's position by adding drift and groundspeed to a previous known position. As we have seen, groundspeed errors are in the order of 0.1%, so along track errors are small. However, compass system errors, or more commonly errors in actually flying the required heading, generate a much larger cumulative cross-track error, even though the drift errors are small. The system error in a Doppler derived position is accepted as being 1% of the total distance flown, and 1° of drift.

Flight deck equipment

Drift and groundspeed indicators

These may be by pointer type indicators or digitally on conventional dials or CRT or LCD displays. A drift and groundspeed indicator, as shown in Figure 13.5(a) below, displays the basic Doppler information.

(a)

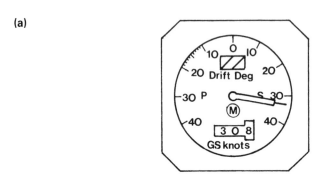

(b)

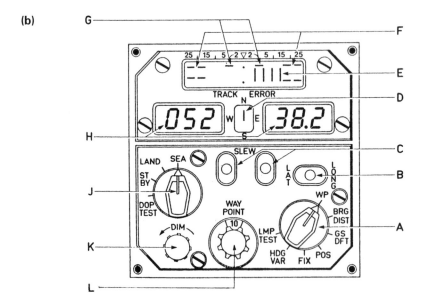

A: Display switch. B: Lat/long selection switch. C: Slew switches (centre-biased and used to update data). D: Light bars to indicate E/W variation, N/S lat, E/W long, etc. E: Track error indicator. The more bars on the right the more the heading must be altered to the right to regain the required track. F: Warning bars (flash to draw the pilot's attention). G: Power supply test bars (used with J). H: Numeric displays. J: Function switch. K: Display illumination brilliance control. L: Waypoint selection switch.

Figure 13.5 Indicators.

Position indicators

Figure 13.5(b) shows a unit which can display whatever information is required by the pilot. As selected on the display switch and shown numerically, this may be aircraft position in latitude and longitude (POS), or bearing and distance from one of several pre-selected waypoints (BRG DIST). It may also display groundspeed and drift, or other details. On the function switch is the land/sea selection, as explained above, and a 'standby' or 'DR' position to manually select the memory circuit. The 'slew switches' move the indicators to either make an input of waypoint latitude and longitude, which is the function (WP) selected on the unit shown, presently 52° 38.2′ N, or to reset the computer to a known position (FIX).

Navigation management computers

Figure 13.6 shows the control and display unit (CDU) of a Doppler navigation computer. The same basic information is available, but more can be displayed, and the increased capacity of a modern computer allows complete routes to be stored in the memory. External inputs can be allowed

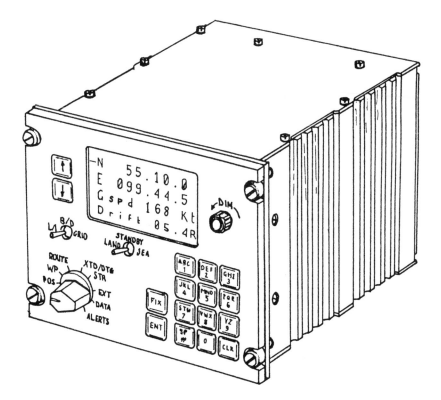

Figure 13.6 Doppler navigation computer.

to appear on the display, and update it without manual slewing. Doppler is a self-contained navigation aid, and as such can form part of an area navigation (RNav) system as described in Chapter 16.

Summary

Principle	Continuous wave – moving target – frequency changed at receiver
	Change in frequency proportional to closing speed
Frequency	8.8 GHz or 13.5 GHz (SHF)
Angle of beam	Depression angle large – small Doppler shift
	Depression angle small – large energy loss by reflection
Measurement	4 beams, rotate aerials till pairs have equal shift, null gives drift
	Shift gives groundspeed
	Fixed beams use computer instead, 3 beams – 2 in front, 1 behind on the left
Janus array	Compensates for attitude, frequency wander
Doppler shift	Proportional to closing speed
	Formula – $df = 2V/\lambda$ df = total Doppler shift, V = closing speed
Accuracy	G/S – 0.1%
	Drift – 0.1°
	Maximum on 95% of occasions = 0.5% of G/S or drift measurement
Errors	• Sea movement – moving sea surface = relative movement
	• Aerial alignment – must be fore and aft
	• Vertical speed – climb or descent = movement, Janus minimises
	• Sea bias – nearest part of beam reflects best. Land/sea sw = G/S + 1%
	• Flat calm – no returns, memory circuit cuts in
Doppler derived position	Main error – pilot holding heading
	Accuracy – 1% of distance flown, 1° of drift

Sample questions

1. An aircraft uses a single radar beam to measure its groundspeed. Which of the following is true?

    a. Groundspeed is proportional to the change in frequency transmitted by the radar
    b. Groundspeed is proportional to the depression angle of the beam
    c. Both (a) and (b)
    d. Neither (a) nor (b)

2. Which of the following is incorrect if the depression angle of a Doppler beam is very small?

    a. A large Doppler shift will be received for a given groundspeed
    b. The power of the reflected beam will be low
    c. The equipment will be most useful over the sea
    d. Groundspeed will be nearly directly proportional to the Doppler shift

3. An aircraft using a single beam Doppler system to measure its groundspeed is travelling at 300 m/s. Depression angle is 60°. Transmission frequency is 10 GHz. What will be the Doppler shift?

    a. 20 kHz
    b. 10 kHz
    c. 5 kHz
    d. 2500 Hz

4. An aircraft uses a single beam Doppler system with depression angle 60° to measure its groundspeed. Transmission frequency is 10 GHz. What will be the groundspeed if the Doppler shift is 5 kHz?

    a. 150 kt
    b. 300 kt
    c. 600 kt
    d. 1200 kt

5. How is the antenna arranged in a 3-beam Janus array, using the clock code in relation to the aircraft's heading?

    a. One beam at 12 o'clock, one at 6 o'clock, one at 3 or 9 o'clock
    b. One beam at 12 o'clock, one at 5 o'clock, one at 7 o'clock
    c. One beam at 6 o'clock, one at 1 o'clock, one at 11 o'clock
    d. One beam at 1 o'clock, one at 11 o'clock, one at 5 or 7 o'clock

6. Which of the following is not an advantage of a Janus array, compared with a single beam Doppler system?

    a. Drift can be calculated
    b. Groundspeed calculations are more accurate
    c. Pitch errors are totally eliminated
    d. The effects of frequency variations are reduced

7. An aircraft with a 4-beam Janus array has a depression angle of 60° and an angle of 30° horizontally between the beams and the aircraft longitudinal axis. What will be the Doppler shift in the array if the transmission frequency is 11 500 MHz and the groundspeed is 600 kt?

    a. 2500 Hz
    b. 5 kHz
    c. 10 kHz
    d. 20 kHz

8. Why would the antenna in a 4-beam Janus array rotate until the Doppler shifts between the pairs of antennas are equal?

    a. To find drift
    b. To produce an accurate groundspeed
    c. Both (a) and (b)
    d. Neither (a) nor (b)

9. What does the 'land/sea' switch on a Doppler system do?

    a. It reduces sea movement error
    b. It increases the calculated groundspeed by 1%
    c. It switches in the memory circuit when flying over calm sea
    d. It compensates for coastal effect

10. What is the effective accuracy of a Doppler groundspeed indication?

    a. ± 1°
    b. ± 1%
    c. ± 0.5%
    d. ± 0.1%

11. What is the effective accuracy of a Doppler derived position?

    a. ± 0.1% of distance flown, ± 0.1° of drift
    b. ± 0.5% of distance flown, ± 0.5° of drift
    c. ± 1% of distance flown, ± 1° of drift
    d. ± 1.5% of distance flown, ± 1.5° of drift

12. Which wavelength would you expect a Doppler groundspeed and drift indicator to use?

    a. 3.5 cm
    b. 35 cm
    c. 3.5 m
    d. 3500 m

13. What is the main cause of error when using Doppler to find your position?

    a. Pitch error
    b. Aerial alignment
    c. Sea movement error
    d. Variations in aircraft heading

Chapter 14
Secondary Radar Theory and DME

Introduction

During the Second World War, the military forces using primary radar found they were unable to distinguish between friendly and enemy targets on a radar screen. To reduce the risk of intercepting and perhaps destroying friendly aircraft, a system of 'identification friend or foe' (IFF) was developed. When a friendly target received a radar signal, it would broadcast its own signal, which could be received by the originating radar receiver. By coding this 'secondary' signal, the aircraft equipment could indicate to the operator of the originating radar that it was friendly.

An advantage of this system was that the re-broadcasting or 'transponding' (transmitting responding) targets could be detected at longer ranges. As described in Chapter 9, the strength of reflected signal received at a primary radar antenna is inversely proportional to the fourth power of the range. However, the received signal at the original target is inversely proportional to only the square of the range, as is the strength of the signal broadcast by the target which is received at the 'interrogating' radar.

The power requirements to provide adequate returns at long range while using a narrow beam width and short pulse width for target differentiation led to the development of similar secondary radar techniques for civilian use. Detailed information on this secondary surveillance radar (SSR) is contained in Chapter 17. The same secondary radar principle is used for Distance Measuring Equipment (DME), which we shall discuss in this chapter.

Secondary radar principle

A receiver in the transponding equipment is tuned to receive pulse modulated signals at a certain frequency. After a short time delay, the 'transponder' sends its own signal on a slightly different frequency to be received at the original transmitter. The signal from the transponder can be coded as a series of pulses, giving information to the interrogator. Range between the target and the interrogator can be calculated by subtracting the programmed time delay from the total time between transmission of the first pulse and reception of the first reply, and multiplying by the speed of radio waves.

DME

Overview

Transponders are placed at ground stations, which are usually co-located with VOR beacons. Each station has a unique (within the UHF line-of-sight receiving area) receiving frequency in the UHF band between 962 and 1213 MHz. The pilot of an aircraft wishing to find its range from the station must select that frequency on his airborne equipment, which will send a series of pairs of pulses at random time intervals on that frequency to the station. The station transponder replies to every pulse it receives (after a fixed delay of 50 μs) with a another pulse at a frequency 63 MHz removed from the received signal. Each pair of pulses is separated by 12 μs. The emission pattern of both signals is designated P0N.

The aircraft transmitter 'remembers' the sequence of differing time differences between its transmitted pulse pairs and awaits a corresponding pattern of transponded pulse pairs on the different frequency. Once the pattern has been received, the computer in the airborne equipment measures the time from transmission to reception, and calculates the range.

In fact, the frequency on which the equipment operates is not usually published as such. It is either identified as a channel number, such as 46X, or more commonly the VHF frequency which is, or would be, used by a co-located VOR beacon.

Ground equipment

The transponder is constantly replying to whatever signals it detects. If there are no genuine aircraft generated signals, it replies to random noise. The transponder filters the received signals and replies to the strongest it receives, whether generated by an aircraft or by noise. In theory, the closest interrogating aircraft will have their signals replied to. The transponder is able to respond to approximately 2700 pulse pairs every second.

Because the transponding must be omnidirectional, the DME ground aerial is a single pole antenna, cut to the ideal length for 1090 MHz. When co-located with VOR, it is placed on top of the VOR antenna, as in Figure 14.1.

DME ground equipments are designated as 'X' or 'Y' channels. 'X' channels are the most common and have the characteristics described above. The transponded signal is 63 MHz higher than the received signal. 'Y' stations reply at 63 MHz below the received signal, and the pulse pairs are

Figure 14.1   DME antenna on a conventional VOR antenna.

differently spaced. 'X' stations are paired with VOR beacon frequencies with whole number decimals (e.g. 114.30), whereas 'Y' beacons are paired with frequencies using halved decimals (e.g. 114.35).

Identification

The ground station also transmits an identification signal on the same frequency, approximately every 30 seconds. During the identification period, the transponded pulses are replaced by regularly spaced pulses, keyed with the beacon identification letters.

TACAN

A military navigation aid called 'TACAN' (tactical air navigation) uses a DME station to provide its range facility. Modulations in its transponded signal provide bearing information to suitably equipped aircraft in a similar fashion to VOR, but these cannot be interpreted by civilian airborne equipment. The range facilities of TACAN stations can, however, be used as normal DME stations by civilian equipment by selection of the appropriate corresponding VHF frequency. Although 'X' channels are usual in TACAN stations, 'Y' channels are also used, and more frequently than in 'pure' DME stations.

Symbology

On aviation charts, a DME station is represented by a rectangle around the position, thus

a TACAN station is represented thus

a co-located VOR beacon and DME station thus

and a VORTAC (co-located VOR and TACAN) thus

Airborne equipment

In fact the airborne equipment automatically and progressively 'scans' the timebase of its own radar receiver from a zero range time delay to its normal maximum range of 200 nm. It is looking for its own random prf pattern which it is transmitting at 150 pulses per second (pps). Once it finds that pattern, it 'locks on' to that time delay, and displays it as a range.

Drum mechanisms on older equipment may be seen rotating in accordance with that scanning until the signal is locked on. Once locked on, the transmission rate drops to between 24 and 30 pps. Small changes in range caused by the aircraft's movement are followed by the equipment. An 'echo protection' circuit prevents reflections from confusing the equipment by only accepting the first signal at the receiver, which will have travelled by the shortest path and therefore will not have been reflected.

The equipment has a memory circuit which continues to indicate the range even if the signal is not received. In older equipment, this continues for up to 10 s. After that time the search pattern is resumed. This memory is designed to allow for times when the aircraft antenna is masked by the structure during turns. More recent equipment may have a longer memory.

Beacon saturation

Theoretically 126 aircraft can use each station at one time, if none of them are scanning at the high rate. On average, however, we assume that some aircraft will be attempting to 'lock on', so a figure of 100 aircraft is commonly accepted as the average number which can use each station simultaneously. However, some RNAV systems (see Chapter 16) alternate their DME interrogations between several stations, requiring to re-lock every minute or so, which means they are interrogating at the high rate for a considerable period of the time. The ground stations will only reply to the 2700 strongest signals every second, so a large number of high-rate interrogations means fewer aircraft can use each station.

Another complication is that some modern airborne equipment also transmits at higher power, not only 'swamping' the station but 'pushing out' some older, closer, airborne equipment which has a lower transmitter power. 'Beacon saturation' is the name given to the apparent error caused by stronger signals preventing an aircraft receiving a range.

Range measurement

The airborne equipment subtracts the $50\,\mu s$ delay from the time between transmission and reception and displays that time as an equivalent range. This can be calculated by dividing the time by 2 and multiplying by the speed of radio wave propagation C, as may be asked in an examination question. The DME indicator shows the aircraft range from the station, so the pilot knows his aircraft is somewhere on the circumference of a circle with the beacon at the centre and the indicated range as its radius. DME therefore gives a circular position line, which can be combined with one or more other position lines to give a navigational fix.

The accuracy of a position line, as required by ICAO at least 95% of the time, is $\pm\,0.5\,nm$, or $\pm\,3\%$ of the aircraft's range if greater. In fact, DME is the most accurate of the 'classic' navigation aids, which makes it the preferred input to area navigation systems (see Chapter 16).

Assuming there is no beacon saturation, the maximum range is limited by the UHF formula. Most airborne equipment indicates a maximum of 200 nm, but some continue out to 300 nm. As readers can calculate, at that range the interrogating aircraft would need to be at an altitude of about 56 000 ft. At the time of writing, Concorde is the only commercial aircraft type capable of cruising at such altitudes.

Slant range

The ranges calculated are line-of-sight ranges, or 'slant' ranges. As shown in Figure 14.2, the ground range may be considerably less than the indi-cated range. Pythagoras' theorem can be used to calculate the actual ground range if given the aircraft height and indicated range, but for prac-tical purposes at short ranges it is worth remembering the proportions of classic right-angle triangles such as 3:4:5 or 5:12:13 in miles, as shown in the figure.

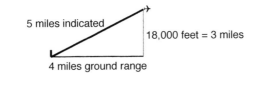

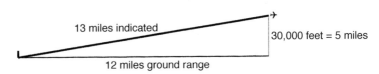

Figure 14.2 Slant range.

Co-located VOR and DME

Most DME stations are co-located, or 'paired' with VOR beacons, as in Figure 14.1. These can be identified because they both have the same published VHF frequency and identification signal. In this case, using the VOR and DME together will produce a pair of position lines (the radius and circumference of the circle centred at the station) which meet at right angles and give simultaneous range and bearing navigation information from the combined VOR/DME station.

Other DME stations may be located close to VOR beacons, but not co-located. These can be recognised by having the same published frequency, but a slightly different identification, in that the DME identification will end in 'Z' instead of the last letter of the nearby VOR beacon's identification. Because the two are not co-located, care must be taken when plotting navigation position lines. At long ranges, however, the information can be regarded for practical purposes as coming from a common source.

ILS paired DME

A DME which is 'paired' with an ILS is designed to give accurate ranges from touchdown along the ILS centreline, as stated in Chapter 8. The DME station of course cannot be situated at the touchdown point itself. The desired result is achieved by altering the transponder fixed delay so that the airborne equipment indicates an incorrect range from the actual station, but the correct distance along the centreline from touchdown. This means that a DME paired with an ILS will give incorrect ranges in any other direction. The DME published VHF frequency is usually the ILS frequency.

Use of the equipment

The pilot must first tune his airborne equipment at the control unit to the designated VHF frequency published for the DME station, then identify the station aurally. Thereafter he need only read the range as indicated on the display. Some equipments offer the ability to select a groundspeed function. This will only work correctly if the station is below the aircraft's track or intended track, in other words if the aircraft is flying towards or away from the station. An extension of that is a time function, which computes and displays the elapsed time to the station if the current groundspeed is maintained. A stand-alone DME display unit is shown at Figure 14.3.

Figure 14.3   DME display.

Range arc tracking

Certain instrument approach procedures include a portion of 'arcing'. This involves following a DME arc at the nominated range until arriving on a VOR radial. In order to reach such an arc prior to following it, the pilot would normally fly along a designated radial towards a co-located VOR/DME. As range reduces towards that designated, he must anticipate reaching it, and start a turn in the required direction which will bring him on to the designated range after 90°. His turn must therefore be started at the designated range plus whatever is the radius of turn.

A 90° turn would take him at a tangent to the radial, and therefore away from the DME station, so his range would increase. The turn should therefore be through only 80°, so that he actually flies along a chord inside the required arc by a small distance. He should then maintain that heading (actually track) until the range displayed increases again to a little above that nominated (usually just under one mile), then turn towards the station to follow a further chord, as in Figure 14.4 below.

On the RMI, while following the chord when the range is just above that nominated for the arc, the VOR pointer will appear a few degrees to one side of (behind) the 90° mark on the instrument. The pilot can use that RMI

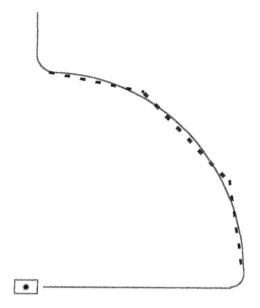

Figure 14.4   Range arc tracking.

needle to help him with the procedure if he wishes. Having noted that angle between the VOR needle and the 90° mark, he should then turn towards the station until the VOR radial is the same amount on the other side of (above) that 90° mark. In this way, as in the technique above, the aircraft will follow a series of chords almost corresponding to the arc of constant range.

Obviously this method does not follow the published procedure exactly, but forms a practical method of flying it. Anticipation is again required before turning on to the final radial, and the 1 in 60 rule can be used to determine that anticipation. Wind will affect the headings to be flown, and should be taken into account to follow the procedure as accurately as possible.

DME summary

Principle	• Secondary radar, a/c interrogates, station transponds, time = range
	• Transmitter prf jitters to recognise own signal
	• Station always transmitting 2700 pulse pairs per second
	• 2700 highest amplitude signals are responded to
Emission	P0N
Frequency	UHF, 962 to 1213 MHz. Transponder changes freq by 63 MHz

Coverage	VHF formula, 'slant' range to 200 nm, Pythagoras for ground range
TACAN	Range part of TACAN is exactly the same as DME
Accuracy	0.5 nm or 3%, whichever is greater.
Errors	Beacon saturation = 100 aircraft in theory, search mode means less
Saturation	• Each aircraft transmits 150 pulse pairs/s in search mode • Only 24–30 pairs/s in normal mode • Switching between stations causes loss of ranges to distant aircraft

Sample questions

1. What accuracy would you expect from a DME range indication of 50 nm?

    a. 0.5 nm
    b. 1 nm
    c. 1.5 nm
    d. 2 nm

2. Under normal circumstances, a DME station can transmit ranges to:

    a. 50 aircraft locked on to it?
    b. 100 aircraft locked on to it?
    c. 50 aircraft in search mode?
    d. 100 aircraft in search mode?

3. If your DME indicator reads 5 nm, and your aircraft is flying at 24 000 ft, what is your surface range from the station?

    a. 3 km
    b. 4 km
    c. 5.5 km
    d. 8 km

4. What form of emission does a DME signal use?

    a. A2A
    b. A3E
    c. A8W
    d. P0N

5. Which of the following is correct?

    a. A DME transponder in an aircraft receives TACAN ranges

    b. A TACAN transponder on the ground sends DME ranges

    c. A VORTAC transmitter on the ground sends TACAN radials

    d. TACAN is not compatible with DME

6. If a DME is paired with an ILS, where will the indicated ranges be correct?

    a. In all directions, up to 20 nm

    b. Within the localiser catchment area

    c. In all directions, up to 10 nm

    d. Along the runway centreline in both directions, up to 20 nm

7. Which of the following is correct?

    a. DME equipment can receive TACAN ranges without modification

    b. A DME station can give ranges to up to 200 aircraft at a time

    c. Simultaneous DME ranges from 2 stations gives a unique fix

    d. The DME station sends a signal 36 MHz different from the interrogator

8. What approximate frequency is used for DME?

    a. 10 GHz

    b. 1000 MHz

    c. 10 MHz

    d. 1000 kHz

Chapter 15
Satellite Navigation Systems

Introduction

In recent years, satellite navigation systems have revolutionised the concept of instrument navigation. Developed as military systems by the United States and the Soviet Union, coupled with the computer systems which have developed at the same time, they allow much more accurate and available position fixing for aircrew. ICAO refers to them as 'global navigation satellite systems' or 'GNSS'.

The United States system, called 'Global Positioning System', or GPS, is the most commonly available. However, the Soviet, now Russian, 'Glonass' system is also commercially available with suitable receiver equipment. Receivers have been produced which can accept signals from both systems, and other countries are considering developing new systems either on their own or in conjunction with others.

Principles of operation

Basic principle

Basically, all satellite navigation systems use the same principle as DME. The receivers measure the time it takes for a radio signal to travel from a transmitter in a satellite at a known point in space. Knowing the time, and the speed of propagation of the waves, a distance can be calculated by the receiver's computer. The receiver then measures the time for a signal from another satellite at a known position to reach it and calculates that distance.

The possible positions from the first satellite can be plotted as a sphere, centred on the satellite, whose radius is the calculated distance. The same applies to the distance from the second satellite, and where the two spheres touch provide the possible positions of the receiver. The two spheres touch along a circle in space, as in Figure 15.1.

A third satellite can provide a further sphere. The possible positions lie on the circle joining spheres 1 and 2, that joining 1 and 3, and also that joining 2 and 3. These three circles meet at only two points in space. One of these is so far removed from the earth that the other must be the correct position of the

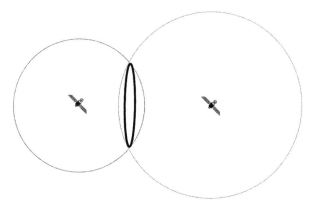

Figure 15.1   Position spheres.

receiver. The computer in the receiver makes all these calculations and displays the result as a three-dimensional point in space related to fixed features on the earth, commonly the lines of latitude and longitude for horizontal, and the centre of the earth for vertical, reference.

Even with only two satellites providing ranges, an observer at a known altitude could feed that altitude into his receiver's computer. That altitude is another sphere in space, so where the two spheres from the satellites touch that altitude sphere is his position. There are of course two possible positions, but if the observer can give his receiver a rough position it will discard the obvious false one. This facility has limited use in aviation, because fortunately there are enough satellites to provide sufficient spheres for three-dimensional position fixing over the whole of the earth for most of the time.

Detailed principle

This simplistic view of the principles of operation has skated over the massive problems of providing the positions of the satellites and the time of transmission of the signals. In order to provide cover over the whole world, the satellites must orbit at an angle to the equator, and so must move quickly in relation to the earth. The transmission from every satellite includes the orbit that each satellite is following, and its position in the orbit, as an 'almanac'. The receiving computer uses that almanac to calculate where each satellite should be in its orbit, so it knows which ones will be in view and what approximate range to expect from each one. This allows faster acquisition of each satellite's signal, and selection of the most appropriate satellites for accurate fixing.

In fact, the ranges as received at the receivers are not exactly accurate.

They are called 'pseudo-ranges', and more calculations must be made to measure the true range from each satellite. The satellite transmits its own exact position and path, called the 'ephemeris', to provide the exact spot from which the ranges must be measured.

To provide correct timing, each satellite carries a very accurate atomic clock. It also transmits, as part of the ephemeris, the corrections which its own clock requires to bring it into line with UTC. To reduce cost, the clocks in the receivers are less accurate. However, by using four satellite signals to compute its position, a receiver will find that the calculated positions between each group of three do not match, as in the complex diagram at Figure 15.2. The error is in the timing, and the receiver computer makes changes to its clock to bring all the computed solutions together. When that happens, its clock is in agreement with those in the satellites. It can then measure true ranges from the satellites instead of pseudo-ranges.

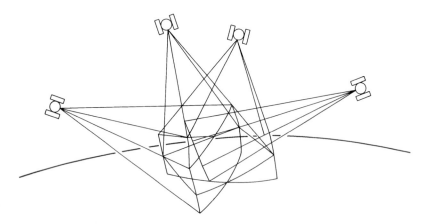

Figure 15.2   Position fixing.

These ranges can be plotted (or computed) to find the receiver's position in relation to the centre of the earth, as Cartesian x, y, and z co-ordinates from that central point. However, navigators traditionally need to know their position in relation to the surface of the earth. The earth is not a sphere, in fact not even a regular spheroid, so the Cartesian positions are compared using mathematical formulae to a theoretical ellipsoid computer model which fairly closely resembles the actual surface of the earth. This allows the receiver's three-dimensional position to be presented in latitude, longitude, and altitude relative to that ellipsoid.

The satellites are supported by ground stations, which receive signals from all the satellites in turn as they pass overhead. The ground stations monitor the signals, and amend the almanac and ephemeris messages transmitted from the satellites.

GPS

Space segment

The American Global Positioning System (GPS), also called NAVSTAR, uses 24 satellites, with three or more spares, in orbit at any one time. The satellites 'fly' in six different orbits, 20 200 km (10 900 nm) above the earth's surface, taking 12 hours for each orbit. The orbits are at 55° to the earth's spin axis, to give the best combination of good coverage, constantly varying geometry, and ease of orbital insertion. All these satellites transmit navigation information on the same frequencies consecutively, in one of 40 time segments or PRNs (pseudo-random noise numbers), but each sends a different code at the start of its transmission. The satellites are usually referred to by the PRN which they use.

At the time of writing, there are two frequencies used, both in the 'L' radar band of the UHF band, but a third frequency is expected to be available for civilian use from 2015. Information between satellites is transmitted on a different frequency, and an S band link carries information from the ground stations (control segment).

Because the satellites are moving, the Doppler effect, described in Chapter 13, changes the received frequencies. The receiver computers compensate by altering their receiver frequencies, using the satellite ephemeris to make the calculations.

The first frequency (L_1), transmitted at 1575.42 MHz, carries the 'coarse acquisition' or 'CA' code, which can be received and interpreted by all receivers. Information carried by the CA code provides the 'standard positioning service', or SPS. The frequency also carries, at 90° phase difference, a 'precision guidance', or 'P' code, which can only be interpreted by military receivers. This 'P' code is also carried on a second frequency (L_2) of 1227.6 MHz, providing the 'precise positioning service' or PPS. The 'P' code can also be 'encrypted' to prevent unauthorised receivers from using it, and when that happens it is referred to as 'Y' code. The third civilian frequency will be called L_5, and will also carry the CA code.

Each satellite carries four clocks, which can be compared internally with each other to provide an average time with a high probability of accuracy. Each one is capable of nanosecond (10^{-9} second) accuracy. However, if one clock is noticeably different from the rest, the satellite can discard that and average the other three. The satellites also carry solar panels and batteries to provide power for their transmitters and receivers, and a system of spinning reaction wheels to keep them pointing at the earth.

GPS satellites have improved progressively. They must be replaced periodically, about every 10 years. Block IIR satellites can provide navigational data without uploads from the ground segment for 180 days, as against the 14 days for the original satellites.

Control segment

The control segment consists of the ground stations. The master station does all the calculations, and the other ground stations provide communications between the satellites and the master station. The master station with its extremely accurate atomic clock compares each satellite's average clock time with its own. It then transmits all the clock errors for re-transmission by every satellite as part of its ephemeris message. The master station can also transmit necessary corrections to the satellite's on-board computer.

If any satellite is transmitting false information, or is in the wrong orbit, the master station tells every satellite that that particular one is unservice-able. Each satellite's transmissions contain information about its own and every other's health, clock corrections, and ephemeris, so the unservice-ability will be re-transmitted in that message. However, it may take several orbits for the unserviceable satellite to fly over the master station after its failure, and the other satellites also have to pass over the master for the information to reach them.

User segment (aircraft equipment)

The GPS receiver is a simple device, requiring little more than a small screened aerial mounted on the skin of the aircraft. Hand-held devices are capable of showing position to previously unimagined accuracy. The computers and their associated software are the complex parts of the aircraft equipment. Computers can calculate actual track and groundspeed made good, and required track and ETA to intended waypoints. Given inputs from flight instruments or an air data computer, the GPS computers can make all navigational calculations. Data loaded into the computer software can display instructions to follow procedures, and direct the aircraft clear of restricted airspace.

A receiver which has no knowledge of the positions of the satellites must download a full set of almanac information before it can receive navigation signals. When first switched on, the receiver will search for any signals on the correct frequencies. Once the almanac message is received from any satellite, the receiver must load its own computer with the information. Once it has received the almanac from one satellite, the receiver selects one satellite whose signal it should be able to receive. It generates its own version of the pseudo-random code being transmitted by that satellite, and searches for the same code at approximately the correct time. Once it has matched the codes, it 'locks on' to that signal. It then looks for signals from the other satellites it expects to receive.

The receiver itself can be one of three categories; multi-channel, multiplex, or sequential. Sequential receivers lock on to one satellite at a time, measure the pseudo-range, then search and lock on to the next. This is of little use in aviation, where the receiver is moving quickly. Multi-channel receivers are expensive, because they have to receive the signals from all the satellites in view independently and simultaneously, but are the preferred type for aviation.

Multiplex receivers 'time share' their receiving and computing time between all the satellites in view. Because the satellites send their information at different times in the short transmission sequence (different prns), a multiplex receiver can store the information from each while it is receiving information from the others. Once the receiver has 'locked on' to a particular satellite, the computer can process the information from it during the time it would normally be attempting to find its signal. When another satellite's signal is due, the computer stores the information from the first to work on it again later in the cycle. It therefore progressively builds up its navigation information.

Whatever system the receiver uses, the time between first switching it on to being able to compute an accurate position, the 'time to first fix', is quite long. Figures of between 10 and 20 min are quoted for different equipments. This time can be reduced if the receiver knows its own position either from having been switched off there, or having the correct information put into it by the operator. It can be reduced even more if the almanac which it last used is still current, so that it can select the satellites without waiting to download the new almanac.

Once continuous navigation signals have been received from four satellites, the receiver computer's first task is to synchronise its own clock (actually a quartz oscillator), by computing fixes from each group of three and comparing them, as explained earlier. This eliminates its own clock error, and allows a final accurate position to be calculated. Several aircraft clocks are available with GPS inputs to show this accurate time to the pilot.

Pilot interface with some GPS instruments is by means of a numeric keypad, but those designed for light aircraft tend to use a menu system. A full computer or alpha-numeric keyboard controls the flight management system of airliners, and is used to interface with the GPS equipment. Functions accessible allow loading of waypoints, often in three dimensions, arranging flight profiles by listing the desired waypoints in the order required, and the 'go-to' or 'direct' function, which with one keystroke gives directions to any waypoint selected. 'Go-to' is an invaluable function, giving instant directions for diversion if necessary. However, it must be used with care, as without an up-to-date database loaded, it may lead pilots through restricted airspace or controlled airspace without clearance.

Figure 15.3 shows a typical GPS control and display unit (CDU) available for light aircraft, showing the 'go-to' key marked with a D with an arrow through it, and a button marked 'NRST' to select the nearest aerodrome from the database (for ease of diversion calculations in an emergency). The information on the screen includes the destination on this leg (the mode selected), with distance to go and groundspeed. It also shows a pictorial representation of the flight path, with the boundaries of controlled or restricted airspace to the right of the track. Listed below the screen are the menus available, controlled by the knob on the right and the cursor (CRSR) switch.

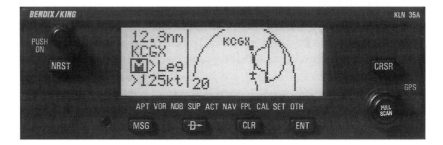

Figure 15.3 Light aircraft GPS unit (courtesy Honeywell).

Figure 15.4 shows a more sophisticated CDU with a colour screen which has been developed for business aircraft. The database includes a map of the ground below the aircraft's track, complete with contours and current aviation database. Although this is a stand-alone CDU primarily dedicated to GPS, inputs from other equipment, such as lightning detection (see Chapter 12) can be displayed on the same screen, and a similar 'moving map' can be displayed on an EFIS screen with inputs from an area navigation system, as described in Chapter 16. The control at the top right of the screen is a 'joystick' for controlling the cursor and scrolling through the menus.

- Position dilution of position (PDOP) is the total effect of all the errors in three dimensions, latitude, longitude, and altitude.
- Horizontal dilution of precision (HDOP) is the total effect in the earth's horizontal plane (distance)
- Vertical dilution of precision (VDOP) is the total effect in the earth's vertical plane (altitude)
- Time dilution of precision (TDOP) is the effect the errors have on the accuracy of the receiver's clock

Figure 15.4 Colour GPS display (courtesy Honeywell).

- User equivalent range error (UERE) is the effect the errors have on the accuracy of calculating the range to one satellite

Selective availability

The 'CA' code provides accuracy in the region of 30 m. However, the US military considered that accuracy too great for use by possibly unfriendly nations. They have the facility to place an irregular (but known to them) time delay in the transmissions. This is called 'selective availability', and historically reduced the 95% accuracy of a GPS fix to 100 m. When GPS was originally made available for civilian use, selective availability was permanently in force, but in May 2000 it was discontinued. The US government has, however, reserved the right to re-impose it for national security, and also to switch off or even jam the signals if it so wishes.

Satellite position error

The orbit of each satellite, and its position in that orbit, is affected by gravity. The gravitational pull comes not only from the earth, but also from the moon, the sun, and the planets. All of these are predictable, and computer calculations to correct these have all been made, but other objects may also have an effect. Other satellites, space debris, and even passing asteroids will all affect the ephemeris to some small degree. Minor malfunctions in the satellites themselves may produce changes in either the orbit or the clocks. As mentioned before, there is a delay before the master ground station can correct, or broadcast the errors in, satellites' ephemeres. Receivers may therefore be calculating positions from incorrect data.

Satellite clock error

The accuracy of the satellite clocks has been mentioned above. However, even the tiny errors which the satellite clocks suffer, despite their regular updates from the master station, can affect the calculated position by about 1.5 metres.

NOTAMS

If a satellite is taken out of service, for example for servicing, a GPS NOTAM will be issued. That will refer to the particular satellite by its pseudo-random noise (PRN) number, then give the date and times between which it is unusable. During these times, pilots should 'de-select' the particular satellite from their navigation computers if that function is available, in order to minimise time wasted by the equipment searching for a non-existent or useless signal.

Ionospheric refraction

UHF signals are not normally regarded as being refracted by the ionosphere, but such accuracy is required that even the very small amount of refraction they suffer increases the time taken for the signal to reach the receiver as it bends through a shallow angle. This can be seen in Figure 15.5. The total distance covered by the signal along the solid line in the figure is greater than the correct path shown by the dashed line. It is especially noticeable when the signals pass through the layers of the ionosphere at a shallow angle, and is less of a problem when the satellite is overhead the receiver. The signal is also subject to attenuation as it passes through

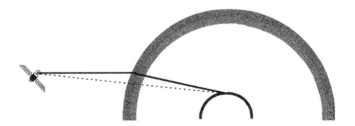

Figure 15.5 Ionospheric group delay.

the layer of ions, which is effectively thicker and therefore has more effect at a shallow angle.

The combined error is called the 'ionospheric group delay' which when combined with the delay from other satellites produces a total position error in the order of five metres. In fact, ionospheric group delay is inversely proportional to the square of the frequency, so if the receiver receives two frequencies, as the military ones do, the computer can calculate the correct ranges from the two different ones received.

Tropospheric refraction

Refraction and attenuation also takes place in the troposphere, producing a 'tropospheric group delay'. However, this is small and regarded as an acceptable error for most applications.

Solar disturbances

The ionosphere and the earth's magnetic field protect the surface, and the troposphere, from damaging radiation. However, satellites fly well above the earth's surface and away from that protection. The 'solar wind' carries that radiation, which includes electromagnetic disturbances and particles, as a stream of energy which can cause 'perturbations' to the satellites' orbits, and in extreme cases cause damage to the satellites themselves. Work is being carried out to detect changes in the solar wind before they reach the earth, by positioning warning satellites in orbit between the earth and the sun.

Receiver measurement errors

Like any man-made objects, the receivers have small inaccuracies in their construction which give rise to errors in the position calculated. These are very small, in the order of 0.3 metres.

Geometric dilution of position (GDOP)

Geometry requires that the satellites used for position calculation must be in different parts of the sky. The spheres of constant range described earlier must cut each other 'cleanly', at an angle approximating to an ideal 60°. Receiver computers will use the almanac to decide which are the most advantageous satellites, and use the signals from those only. However, at certain times it may not be possible to find enough satellites giving that clean cut, perhaps because of unserviceability or maintenance. GDOP has been known to produce a large error for about 20 min in any one day.

Multi-path errors

UHF signals are subject to reflection from many surfaces, and a reflected signal will arrive at the receiver later than the direct wave. Usually, the reflected signal will be weaker than the direct wave and the receiver computer can reject it, but where the direct wave is 'masked' by some object in its direct path, the reflected signal may be accepted as the true signal, giving a range error and therefore a position error. Receiver antenna design can reduce multi-path errors by discarding signals which arrive from unusual angles. For example, any signal received from the ground beneath the aircraft, or from the direction of the aircraft's fin structure, should be disregarded. On average, multi-path signals produce a final position error of about 0.6 metres.

Interference

The GPS signals, being broadcast from such a high altitude, are very weak when received. The signal to noise ratio is actually less than one, so the receiver has to identify the weak signal within the noise by its characteristics. As the electromagnetic spectrum becomes more crowded, harmonic interference from other transmissions, perhaps satellite communications transmissions from a nearby transmitter or the aircraft itself, may be relatively strong enough to cause interference. Deliberate interference can be produced in the form of 'jamming' from quite a weak transmitter close to the target receiver, and such jamming equipment has been made available for sale commercially in recent years.

GPS accuracy and integrity

Accuracy

The CA signal (unless subject to selective availability) is assumed to be accurate in the horizontal plane to within 30 m for 95% of the time. It is considered accurate to within 300 m for 99.99% of the time. Because the receiver uses satellites which are high in the sky, the computed vertical position is less accurate. The 95% position should be within 156 m (500 ft) vertically, and the 99.99% position within 500 m (1600 ft).

Carrier wave measurement

Very small changes in position, in the order of centimetres, can be measured by comparing the phase changes of the signal carrier waves. If the satellites were geostationary (in a fixed position above a point on the earth), the changes in phase would equate to movement on the surface of the earth, but the moving satellites require major calculations from the satellite ephemeres to compare the actual phase changes with the changes to be expected from the satellites' orbits. This method is used successfully in survey work. However, at the time of writing, carrier wave measurement is not in use for aviation purposes, although it has been suggested that it could become a source for velocity calculations in the future.

RAIM

Because of the possible errors in the system, regulating authorities demand that, before it can be used as a primary navigation aid, a GPS receiver must be able to detect when its information is unreliable. In other words, it must monitor its own 'integrity'. Using more satellites, this can be achieved. Four signals allow timing corrections, and five signals allow the receiver to detect that one of the signals is incorrect. This is known as 'remote autonomous integrity monitoring', or RAIM. Six signals allow the receiver computer to use a 'failure detection and exclusion, or FDE' algorithm to determine which is the incorrect signal and discard it.

A satellite system used for 'stand-alone' navigation, and especially precision approach guidance, must be capable of monitoring its signals well enough to prevent errors affecting its accuracy. It must also warn the crew of any errors which might affect that accuracy. To do that it needs to know when the required six satellites will be 'in view' or available for reception. The system therefore must predict that availability, and warn the crew in advance of any time when insufficient signals will be available. This is referred to as predicted RAIM (P-RAIM). It is feasible that the receiver computer can use barometric altitude information to act as one of the positioning spheres (in fact an ellipsoid) to allow RAIM with only four satellite signals, or FDE with five.

Other satellite systems

Glonass

The Russian satellite navigation system is called GLONASS. It uses separate frequencies for each of its 21 active satellites (which use a slightly different orbit at 60° to the GPS satellites' 55°) to transmit navigational data. There is no selective availability facility in the system. All Glonass satellites orbit the earth every 12 h at an altitude of around 19 000 km.

European GNSS

The states of the European Union have been working on a system of their own to provide satellite navigation information independent of the United States GPS system. At the time of writing, it is still at the proposal stage. It is intended to use geostationary communications satellites over the equator to relay signals from ground stations which can be received as if they were satellite generated, but using a ground atomic clock for accuracy. The ground stations which generate the signals are known as 'pseudo-lites'.

It is proposed to use the pseudo-lites in conjunction with normal navigation satellites, either GLONASS, GPS or independent. The higher orbit of the geostationary communication satellites which would relay the pseudo-lite signals, coupled with the extra distance the signal would have to travel from the ground station to the relaying satellite, results in a very long delay from transmission to reception of the pseudo-lite signal, which must be taken into consideration by the receiver computer.

Differential GPS

Mainly because of selective availability, scientists looked at ways of increasing the accuracy available from GPS. If a position on the surface of the earth was known exactly, and a GPS receiver was placed at that position, there would be a difference between the GPS position and the true position, caused by all the errors (except receiver measurement error). That difference would be almost exactly the same between any GPS position and a true position in the area of the known point. If the difference could be transmitted to users, their receivers could add the position difference to their indicated GPS position, and they would then know their true position to a much greater degree of accuracy.

In fact, a computer at the known ground position calculates the true ranges (and therefore time delays) from each satellite. It compares that calculated time delay with the time of the received signal, and transmits corrections for each signal to its customers. Computers in customers' receivers can then make the required corrections to the pseudo–ranges they have received to find their true position. This system is known as differential GPS or DGPS. DGPS removes SA and group delay errors, and allows accuracies in the order of three metres. It can also provide external integrity monitoring which can supplement or replace RAIM.

The correction messages can be transmitted by several media, and can be encrypted for sale or available free. In the UK, sub-carriers of FM commercial radio broadcast frequencies are used. On the coast of Europe, NDBs transmit DGPS corrections on their MF frequencies. Surface vessels can use MF signals to find their way into narrow harbours, because they can receive the ground wave. DGPS corrections can also be transmitted through communications satellites. Whatever method is used, regulating authorities require the corrections, and certainly integrity monitoring information, to be available to the pilot almost instantaneously if the system is to be used for precision approach guidance.

In the USA, two separate DGPS systems are under development. WAAS (Wide Area Augmentation System) is designed to give accurate en-route navigation over the continental United States, using LORAN-C transmitters to provide the corrections and geostationary communications satellites to transmit them. It is ultimately intended to also provide category I approach facilities. LAAS (local area augmentation system) has been developed to give accuracy in approach procedures initially to category I but ultimately to category III standard.

In Europe, EGNOS (European Geostationary Navigation Overlay Service) is the differential part of the European GNSS. EGNOS intends to use the geostationary communications satellites to re-transmit signals from ground master stations giving the differential corrections required at the same time as they transmit the pseudo-lite signals themselves.

Use of GNSS

En-route procedures

GNSS equipment is often part of an area navigation system (see Chapter 16). If used in a stand-alone mode, its presentation is either similar to that for an area navigation system, or often that of a moving map. In all cases, the receiver computer is the heart of the airborne equipment. It can display the aircraft's position either as latitude and longitude, or distance and bearing to or from selected waypoints. It can also calculate and display track and groundspeed, and the heading and time to the next waypoint. If the computer's database includes information from the AIP and current amendments (published as AIRAC information), it can provide information on controlled and restricted airspace, and suggested avoiding action. Most moving maps, such as that shown in Figure 15.4, include such information on their database when installed, and an amended database is usually loaded in by tape or disk every 28 days.

Directions to selected waypoints may be given by the GNSS display unit itself in the form of a CDI or digitally. If the equipment is integrated into another display system, the directions may be displayed on a flight director or an attitude direction indicator (ADI). The volume on *Navigation* in this series describes these instruments, and some are shown in Chapter 16 of this book.

Approach procedures

GNSS approaches without differential information are regarded as non-precision approaches. In States where GNSS approaches are permitted (not the UK at the time of writing), various criteria must be satisfied. The equipment computer must be loaded with a database including the way-points for the intended approach and other information included in the current AIRAC cycle (published amendments to the AIP). The pilot must consult the NOTAMs to ensure that all necessary satellites are available, and that RAIM can be maintained during his approach. Vertical positioning must be achieved by use of the pressure altimeter, and not a GNSS vertical position.

The equipment will present the approach pattern as directions to a series of waypoints along the procedure. Normal approach procedures include fixes to be overflown in the order initial approach fix (IAF) to final approach fix (FAF) then to the missed approach point (MAP). If a go-around is required, the pilot must fly the aircraft over the missed approach holding point (MAHP). In GNSS approaches, the fixes to be overflown are designated initial approach waypoint (IAWP), final approach waypoint (FAWP), missed approach waypoint (MAWP) and missed approach holding waypoint (MAHWP) respectively. Any other points to be overflown are referred to as intermediate waypoints (IWP).

The CDI used for GNSS approaches must change its sensitivity during the approach, either automatically or, more usually, as selected by the pilot. Before commencing the approach, full-scale deflection on the CDI represents 5 nm, as in any RNav equipment (see Chapter 16). Once the pilot selects 'approach mode', the full-scale deflection represents 1 nm, and each dot 0.2 nm. Just before the FAWP, the sensitivity changes again, usually progressively, to give a full-scale deflection representing 0.3 nm to one side of the centreline, and some equipment indicates an angular displacement in the same way as ILS or MLS displays.

The use of GPS has been strongly supported by manufacturers and the government in the USA. These supporters suggest that differential GNSS will provide a precision approach system of comparable accuracy to MLS at much less cost. However, precision satellite approaches are still some time in the future. Equipment manufacturers have had to prepare for the arrival of both new systems, and at the same time make improvements to ILS receivers and indicators. At the time of writing, there are multi-mode receivers available which provide the same indications for ILS, MLS, and GPS approaches, and which can be set up in the future for precision approaches using DGPS.

Summary

Principle	• Satellites send their position and a time signal • Receiver compares Tx code with its memory of code • Notes time difference between transmission and reception • Time × speed of light = distance, DME – like 'pseudo-ranges'
Fixing	• 1 satellite gives position sphere • 2 satellites give ambiguous position on earth's surface • 3 satellites give ambiguous position in space • 4 satellites remove time errors and give accurate stand-alone fix • 6 or more provide remote autonomous integrity monitoring (RAIM)
Frequencies	• All GPS satellites use same frequencies, multiplexed and coded • Nav signals in L band (UHF), $L_1 + L_2$ (1575.42/1227.6 MHz) • 3rd frequency L_5 due 2015 (1176.45 MHz) • Frequency at Rx varies with Doppler shift – tuned by Rx computer • Separate UHF freq to other satellites • S band link to ground master station
GPS	• 21 active satellites plus 3 spares • L_1 sends P(recision) code signal – PPS – split into 40 segments (PRNs) • Each satellite has a PRN and code to match • C/A (coarse acquisition) code 90° out of phase of P code in L_1 signal • L_2 signal only sends P code • P code may be further encrypted, called Y code • Almanac and serviceability for all satellites in each signal
Selective availability	– (S/A) may degrade civilian (SPS) signal by inducing time errors
Errors	• Satellite position errors – Clock error and ephemeris error

	• Ionospheric group delay – • depends on angle through layers
	• error depends on $1/f^2$
	• 2 frequencies can find it and reduce it (PPS)
	• Tropospheric group delay – small
	• Multi-path errors – reflections
	• Receiver measurement errors
	• GDOP – geometric dilution of position – up to 20 minutes per day
	• Interference – onboard or external close signal/harmonics/Satcom
Accuracy	• CA signal accurate to 30 m for 95% of time, 300 m for 99.99%
	• Vertical accuracies 500 ft for 95%, 1500 ft for 99.99%
GLONASS	21 active satellites, Russian owned, no degraded signal, separate freqs
DGPS	Differential GPS – known position transmits error corrections
	Removes SA and ionosphere error, accurate to 3 m
	Currently MF (NDBs), also sub-carrier of VHF FM
	WAAS – for en-route
	LAAS – for precision approaches
GNSS 1	• Adds EGNOS (European WAAS) to geostationary satellites
	• INMARSAT transponds ground signals to add to GPS signals

Sample questions

1. How many GPS satellites are needed to provide an accurate stand-alone fix?

   a. 1
   b. 2
   c. 3
   d. 4

2. How many GPS satellites are needed to provide remote autonomous integrity monitoring for precision guidance?

   a. 3
   b. 4
   c. 6
   d. 8

3. How does a GPS receiver determine from which satellites it is receiving the signal?

   a. Each satellite uses its own frequency
   b. The signals are sent on the same frequency and coded
   c. Each satellite sends its signals at different times
   d. The signals are sent on different frequencies and coded

4. Which of the following is not an error of GPS?

   a. Ephemeris error
   b. Geometric dilution of position
   c. Clock error
   d. Tropospheric ducting

5. Which of the following does DGPS not reduce?

   a. Ionospheric group delay
   b. Multi-path errors
   c. Selective availability
   d. Satellite position errors

6. What does a GPS receiver measure?

   a. The time between transmission and reception of a signal from a satellite
   b. The time for a signal to pass from it to the satellite and return
   c. The time between reception of signals from two satellites
   d. The phase difference between signals from two satellites

7. Which GPS signal or code is available for use by civilians?

   a. The L_2 signal
   b. The Y code
   c. The P code
   d. The SPS signal

8. Which of the following is incorrect about the coarse acquisition code of GPS signals?

   a. It is transmitted 90° out of phase from the precision code
   b. It contains the PRN of each individual satellite
   c. It allows an accuracy of ± 30 m on 99% of occasions
   d. It is encrypted for US military use

9. GLONASS is the Russian equivalent of GPS. Which of the following is correct?

   a. 18 GLONASS satellites cover the world
   b. GLONASS can be combined with GPS to reduce GPS GDOP error
   c. The GLONASS signal is degraded by the Russian military
   d. The GLONASS satellites all transmit on the same frequency

10. Which of the following is incorrect about differential GPS?

   a. It reduces the effect of selective availability
   b. It reduces ionospheric error
   c. It reduces geometric dilution of position
   d. It increases accuracy

11. What does a GPS receiver measure to compute its position?

   a. The ranges from the satellites
   b. The difference in time from signal transmission to reception
   c. The Doppler difference in signal speed from transmission to reception
   d. The difference between the satellite clock and the receiver clock

Chapter 16
Area Navigation Systems (RNAV)

Introduction

The historical structure of airways over the world, where aircraft flew from one navigation aid to another, was adequate while air travel was limited. As business and leisure air travel became more popular, and more aircraft were built to use the airspace, the limitations of a system which channelled all the aircraft with their different speeds along fixed routes became obvious. Fortunately, computer technology was able to provide some solutions to these limitations.

As early as the 1960s, area navigation systems had become available for aircraft. At first, these were simple mechanical devices which altered the display from the radio navigation aids to simulate the repositioning of the particular aids on to the aircraft's planned track. The pilot could then fly towards these 'offset' navigation aids. Electronic computation made such devices lighter, and expanded the possibilities of the technique. Crews could produce a flight plan which did not have to pass over any radio aids at all, and the navigation computer would guide them along it.

Eventually, the computers in an aircraft developed to a stage when they could reproduce the aircraft's performance parameters. They could then calculate ideal vertical flight paths for climb and descent to arrive over places at particular altitudes. The Flight Management System (FMS), coupled to the autopilot, can now fly the whole route which the crew have planned in three dimensions. Even the fourth dimension, time, can be controlled by the FMS, as airspeed adjustments can be computed to make required times at certain positions.

VOR/DME RNav

A simple area navigation system uses the received signals from VOR and DME stations with a dedicated computer and a CDI or HSI to guide the pilot along his planned route in two dimensions. The crew select a series of 'waypoints' which are their planned turning points or timing points. Such waypoints may be referred to as 'phantom stations'. To define these phantom stations, the crew must calculate or copy the ranges and bearings of each waypoint from one or more pairs of VOR or DME stations. The computer then calculates and displays directions on the CDI or HSI to the pilot to keep the aircraft on the track between the last waypoint and the next, as if the phantoms were actual VOR/DME stations.

Inputs

As suggested by the title, the inputs are from DMEs and VORs, usually two of each. ADF is not regarded as accurate enough to provide a suitable input to the RNAV computer. The most accurate guidance is provided when the computer is able to fix the aircraft's position using two, or ideally more, DME ranges. This is sometimes referred to as a 'rho-rho' (ρ-ρ) fix. Less accurate is a fix using a VOR radial and a DME range, referred to as a rho-theta (ρ-θ) fix. Least accurate, but acceptable, is a theta-theta (θ-θ) fix from two VOR stations.

Use of RNAV

In a basic system, the pilot must manually select the VOR and DME stations he wants the computer to use, and identify them as normal. When moving outside the range of each aid, he must select, tune and identify the aids he wishes to continue with. He should also monitor his progress continually, using other aids.

Display

The CDI or HSI displays lateral displacement (distance to one side) from planned track. When used in this fashion, the display is slightly changed from the basic VOR display. In RNAV mode, the horizontal dots on the instrument face represent not an angular displacement from the VOR radial, as described in Chapter 6, but a distance to the side of the track. Each dot represents one nautical mile displacement, so the display in Figure 16.1 below indicates that the aircraft is 2 miles to the right of the track of 045°. The DME display indicates distance to go to the next waypoint (23 nm).

Control unit

A control display unit (CDU) incorporates the selectors for the basic system. The pilot must select the range and bearing of the phantom waypoint from each of the inputs (which he has already identified) on the CDU. The unit in turn will display a digital readout of these ranges and bearings, and in addition usually of the track and range to the selected waypoint. A switch provides the facility to prepare the next waypoint while still providing

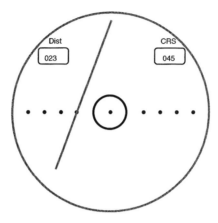

Figure 16.1   RNAV lateral displacement display.

directions to the present one. As technology has advanced, CDUs for this 'short range' RNAV equipment have evolved, and most now bear strong similarities to those in figures shown below.

Benefits of RNAV

When using RNAV equipment, aircraft are not constrained to follow the airway system. Consequently, there are several advantages to be gained by operators of such aircraft. They can plan to fly more direct routes, thereby saving fuel, cost and environmental damage. Fast aircraft may be able to overtake slower ones laterally by changing routes, allowing a smoother flow of traffic. Aircraft departing from aerodromes can be routed away from previous traffic which may be slower or have a lesser rate of climb by directing them to phantom waypoints. Aircraft holding before descent can hold at phantom stations at their ideal descent point rather than congested VOR beacons closer to their destination which may involve additional air miles in the descent.

Accuracy, reliability and coverage

Because the system uses actual VOR and DME stations, the accuracy, reliability and coverage depends on that of the selected real stations. The range of those primary aids defines the cover available. The position accuracy depends not only on the individual accuracies of the basic aids, but also by the geometric positioning of the stations themselves. Accuracy is also affected by the fact that the DME ranges, which are the most accurate and preferred inputs, are slant ranges, and the equipment is not normally set up to make the Pythagoras calculations.

More advanced systems

The availability of other inputs, and the improvement in computer processing power, led to the development of more sophisticated RNAV systems. At the same time, the increased reliability of these other inputs, and of aircraft instruments, have allowed pilots and controllers to reduce separation between aircraft in flight with as much safety as before, or more.

That increased instrument reliability, a consequence of fitting air data computers (ADCs), can be coupled with the performance information in the FMS to allow the RNAV equipment to produce directions in the vertical plane (V-NAV) as well as the horizontal, or lateral (L-NAV).

Inputs

Any accurate navigation aid can provide an input to a RNAV system. However, the main improvement came with the input from one or more inertial navigation systems (INS). This gives a RNAV system a reliable input even outside the range of DME or VOR stations. Omega (see Chapter 23) was a prime input for a while, and Loran-C (see Chapter 22) may be currently used, as well as DME and VOR. Doppler (see Chapter 13) is available for use as a self-contained aid (requiring no external radio station) with the potential to replace one or more of the inertial navigation systems, and GPS (see Chapter 15) provides a major input to most systems.

Air data inputs are also required by modern systems to provide the V-NAV commands. From the Air Data Computer (ADC) comes Mach number, airspeed, temperature, pressure altitude and vertical speed information. These allow the performance parameters to be calculated, and can be compared with the position calculations to provide displays of drift and wind velocity if desired.

Self-contained on-board systems, such as inertial navigation systems, are triplicated in most airliners, to increase the accuracy by averaging out the three positions. Triplication also provides back-up in the event of failure of one or two of them. In addition, the outputs from each INS can be compared by computer, and if one is different from the rest the computer can discard its information automatically and display a failure signal to the pilot.

As explained in the relevant chapters on each aid, the errors of self-contained systems, such as INS or Doppler, are proportional to the distance travelled (an INS position is generally accepted as more accurate than a Doppler position). In contrast, the errors of radio navigation aids such as DME and VOR depend mainly on the range from the transmitting stations. These different types of input with their different error characteristics complement each other. Self-contained systems are said to have very good 'short-term stability', in that their errors are small immediately after switching them on. However, they have poor 'long-term stability', because their errors increase with time and they become progressively more inaccurate.

Inputs from the ground-based radio aids have errors which remain basically constant with time, depending only on the range from the stations. They have poor short-term stability, but provided that such inputs can be received from new stations when the aircraft has travelled a long distance, the effectively constant error has not changed, so they have good long-term stability.

The accuracy of GPS, especially if updated with differential information and checked with RAIM, makes it the manufacturers' preferred external input for many modern RNAV systems. However, its limitations mean that many States do not allow its use as a primary source of navigation information, so other inputs are essential.

RNAV System Components

Navigation computer unit

The navigation computer unit (NCU) is the computer which takes the inputs from navigation aids to provide the lateral navigation (L-NAV) directions. It compares all the available inputs in a system called 'hybrid navigation', and uses a technique called 'Kalman filtering' to arrive at the most probable present position for the aircraft. This Kalman filtering does not just average the various inputs; it gives a calculated 'weighting' to each of them so that the most accurate has more effect on the final position than a less accurate aid.

Most systems separate the position calculated by the self-contained system inputs (either averaged, filtered, or selected individually) from the position calculated by the NCU. On the display of the CDU (see below) the pilot sees the two positions and a vector showing the difference between them. He can then manually accept the NCU position to update the self-contained position if he so desires.

The quality and complexity of the Kalman filtering algorithms have a considerable effect on the accuracy of the final computed position. Even without external inputs with good long-term stability, a good filter can use previously found errors in the self-contained systems to produce an estimated position with a high probability of accuracy. The NCU compares the computed position with the planned track, and produces an indication of the distance to go to the next waypoint, and the distance off track, as a simple RNAV would do. However, the NCU can also produce a L-NAV command to bring the aircraft back on to track, either at the next waypoint or earlier if preferred. This earlier return to track is needed for flight in all airspace requiring enhanced navigation systems.

The crew may have made the aircraft deviate from the original plan, for example to avoid weather or another aircraft. In this case, the NCU will direct the aircraft's return to the original plan either by taking them to the next suitable waypoint or back on to the current track as desired. The output from the NCU can also be fed direct to the autopilot when selected by the pilot.

Flight data storage unit

The modern systems no longer require the pilot to select the navigation inputs he wishes to use (although he may if he so wishes). Nor need he dial in the range and bearing of his phantom waypoints. The flight data storage unit (FDSU) contains the position and protected ranges of all navigation aids throughout the world, or at least along the route to be flown. When the pilot selects his phantom waypoint on the CDU (see below) the NCU selects the stations which will give the most accurate fix from the FDSU.

The FDSU in current systems also contains a wealth of information about aerodromes and positions of points on them, Air Traffic Services routes, company routes, standard arrival routes (STARs), and standard instrument departure procedures (SIDs). It contains information on magnetic variation around the globe, and this can be applied by the NCU when selected by the pilot. The database of aerodromes also includes runway threshold and even parking stand positions. This means that the FMS can be set up to the stand position, and automatically updated as the aircraft starts its take-off run.

Information about aerodromes, reference points, navigation aids and procedures, which are contained in the national AIPs, change and usually requires regular updating. Commercial organisations usually provide the databases with a regular update facility (amendments to the AIPs called AIRACs) on a computer disk or tape which contains all the changes. The updates are required every 28 days, and the next disk is made available early, so that its information can be stored in advance and exchanged for the old information in the FDSU when it becomes effective. On that date on which the new database becomes valid, the crew will change the database information from old to new by selecting it on the flight management system (FMS).

The FDSU also contains the aircraft's performance details to add to the air data inputs to provide V-NAV guidance. Such information includes economic performance as well as maximum performance details, and can even take account of differences in fuel pricing between aerodromes!

Control display unit

The control display unit (CDU) of a modern system has the same function as that in the simple RNAV system. It is an interface between the crew and the computer. The display part can give a digital indication of the aircraft's computed position in latitude and longitude. It can also indicate the position of the next selected waypoint, the distance to run to it, the groundspeed made good along track, and the estimated time of arrival at the next way-point. However, the CDU normally also feeds the navigation display of the EFIS horizontal situation indicator, which leaves the CDU screen free to act as a computer control screen. Figure 16.2 shows a representation of a typical CDU, although some may use a typewriter keyboard layout for the alphabet keys.

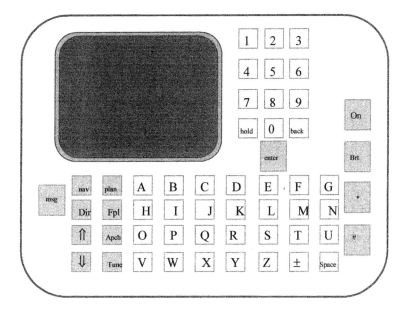

Figure 16.2   CDU.

When the aircraft is fitted with a flight management system (FMS), the NCU forms part of the more extensive flight management computer (FMC). The CDU displays a series of pages of information either provided or required by the FMC. To change the information displayed on one of the data fields on the unit, the pilot writes the details on a 'scratchpad' at the bottom of the CDU screen, then transfers it to the particular field to which it refers by pressing the selection key of that field. Figure 16.3 shows the CDU of a typical FMS, normally placed horizontally on the centre console in the cockpit, aft of the engine controls. As described in Chapter 3, the scratchpad

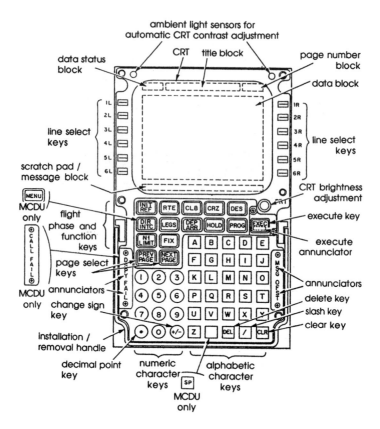

Figure 16.3 CDU of a FMS

is also used to write and receive ACARS messages. These FMS pages contain data on the various phases of flight. One page may show the aircraft's initial position by aerodrome and gate with the mass and balance figures. Another will detail the take-off performance parameters. Others will show, for example, the lateral and vertical positions of the ATC departure clearance, the waypoints on the route, the three-dimensional STAR positions and the approach details.

On an electronic HSI, the display can be expanded to show only the top centre of a full HSI, as in Figure 16.4. The aircraft's position is shown at the centre of the bottom of the screen. Depending on the display mode, which may be selected on a separate display control unit such as shown at Figure 16.5, the aircraft's track or heading is displayed at the top of the display. A heading 'bug' (visible on the display at 036°) can be moved with the heading knob to prepare for any loss of automatic function. The radial or ILS centreline which the aircraft should follow can also be manually selected with the course knob. This provides a facility to deviate from a planned route to follow Air Traffic Control instructions, or to avoid weather.

Figure 16.4   EFIS Nav display (courtesy of Meggitt Avionics).

Below the expanded compass rose, the relative positions of the aircraft, the track, and the next few waypoints are displayed on the map. Depending on the selected display mode, this map is oriented either along track or heading, or relative to true North.

The display unit may also include a complete moving map of the terrain below the aircraft, as a projected image from information held in the FDSU (similar to the display shown in Chapter 15, Figure 15.4). It may also show weather from the AWR and/or lightning detector (see Chapter 12), traffic inputs from ACAS (see Chapter 18), or terrain warning information from TAWS (see Chapter 21).

The control part of the FMS CDU contains the keypad with which the pilot can insert the desired waypoints or other inputs. This may be an abbreviated keypad similar to a telephone, or a full keypad with numerals and all the letters of the alphabet, as shown in Figures 16.2 and 16.3. The waypoints

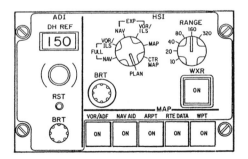

Figure 16.5   Display control unit.

should be inserted in sequence along the planned track, and are usually loaded before flight. This preflight loading may be done either individually by a crew member on the keypad, or as a named route held in the memory of the FDSU and loaded by name. It may also be possible for a datalink system such as ACARS or Mode S to send information to the FMS to insert one or more waypoints or select a pre-loaded route from the FDSU.

The CDU also allows the pilot to load navigation aids not already in the FDSU. More practically, it also allows him to select any navigation aids which he does not want to provide inputs to the Kalman filter algorithm (computer calculation). This is important to prevent a known unreliable aid from affecting what would otherwise be an accurate position. Even an individual satellite can be de-selected if it has been notified as unreliable, or the geometric dilution of position (GDOP – see Chapter 15) is likely to reduce the accuracy of a fix.

During a long flight, the inertial or other self-contained navigation position may be different from the position calculated from the external aids. The CDU may indicate both positions on the navigation map display, and the pilot can decide if and when to re-align the self-contained aid with the computed position.

V-Nav

Using the aircraft performance information stored in the FDSU, possibly as amended by the engine management system, the NCU can calculate the optimum flight level for the particular phase of flight. As fuel and therefore mass reduces during the cruise, the FMS will indicate the best time to change levels if a cruise climb is not practicable. A display showing the vertical profile of the flight, similar to that in Figure 16.6, may be selected on an FMS with this facility.

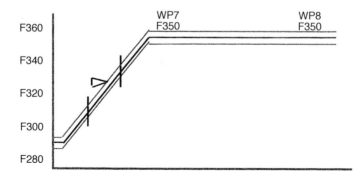

Figure 16.6 FMS vertical display (reversed image).

SIDs

The standard instrument departures (SIDs) from aerodromes are published to ensure the smooth flow of departures. SIDs for all aerodromes which the aircraft is likely to use on particular routes are stored in the FDSU. L-NAV directions from the NCU can guide the aircraft along these SIDs during its climb, reducing the crew's workload.

However, the V-NAV facility allows more sophisticated guidance along the SID. It will direct the crew, or the autopilot more likely, along a three-dimensional tube in space to not only fly over the waypoints which are positioned under the SID path, but also to arrive at each waypoint at the desired altitude, in accordance with any ATC clearance.

In the future, aircraft will be able to follow any departure procedure required by ATC without reference to SIDs. ATC will be able to tell the pilot what positions they want him to overfly, and at what altitude, and the NCU will take him along that three-dimensional route.

STARs

In a similar fashion, aircraft can be directed to follow standard arrival routes (STARs) to guide them on to the runway centreline. Again, the V-NAV facility can direct the aircraft's descent path to be overhead the waypoints at the cleared altitudes, and on to the centreline at an ideal height. Again, future developments will allow any three-dimensional path to be followed. Even without MLS (see Chapter 19), this facility will be able to provide flexibility and efficiency in arrival procedures.

Speed control

At present, speed control tends to be limited to providing adjusted climb or descent parameters. Future developments will allow the equipment to control timing by making speed corrections through the autothrottle. This has been called 'four-dimensional navigation'. The aircraft will be able to maintain desired timing in the cruise or separation on a STAR or SID. This will allow the aircraft to follow precise ATC instructions automatically, effectively giving a controller the power to direct the aircraft as he wishes.

Flight direction

The output from the NCU includes not only the display on the CDU and the EHSI. It also includes a flight director. In its simplest form, this may be a separate zero reader (see Chapter 8, Figure 8.6), but is more likely to be included on the attitude indicator (which is then referred to as an attitude director indicator or ADI). The director information on the ADI may be simple crosshairs like the zero reader, and as shown on Figure 16.7 instructing the pilot to reduce the angle of bank and increase the pitch angle. Alternatively, it may provide something similar to the 'tramline' pitch and bank directions shown on the instrument in Figure 8.7

Figure 16.7   ADI with crosshairs (courtesy Meggitt Avionics).

Autopilot Flight Director System

Most commercial aircraft, and certainly all airliners, are equipped with an autopilot. The NCU can instruct the autopilot to obey its commands through an Autopilot Flight Director System (AFDS). Information from the NCU may be sent independently to the pilot's Flight Director and the AFDS, but usually the same information is required for both, so that the crew can monitor the autopilot's actions. The AFDS controls not only the aircraft's primary flight controls to follow L-NAV commands, but also the engine controls through the auto-throttle to follow V-NAV commands. Just as in manual flight, the pilot can also input commands directly to the AFDS if a change of heading or flight level is required which is not pre-programmed in the FMS.

RNav designated airspace

Parts of European airspace are already specifically designated. Inside that airspace, aircraft are required to carry RNAV equipment which complies with particular specifications. These specifications may be described as 'basic' or 'precision' depending on the accuracy required by the airborne equipment.

Basic RNAV

Equipment complying with the Eurocontrol definition of Basic RNAV or B-RNAV must be accurate to within ± 5 nm in lateral track keeping for 95% of the time. At the time of writing, several B-RNAV routes have been nominated, and on these all aircraft operating at or above FL 100 must carry RNAV equipment which has been proven to meet those standards and approved by their national authority. In addition, they must carry any other navigation aids specified for the route.

Precision RNAV

The accuracies to be expected by Air Traffic Control from Precision RNAV or P-RNAV equipment are more exacting. The quoted figure is ± 1 nm across the intended track for 95% of the time. RNAV airfield approaches require this P-RNAV accuracy, but at the time of writing, no P-RNAV routes or approaches had been designated in Europe.

Summary

Principle	• Computer combination of VORs and DMEs produces waypoints
	• Further computation to give directions to each waypoint
Indications	• CDI and DME indicators show distance along and across track
	• CDI indicates distance off track, not angles
Inputs	• Older systems – DME and VOR (at least two of each)
	• Newer systems – triple self-contained systems (INS, Doppler, GNSS) plus DME and VOR (usually three of each)
Accuracy	• INS (and Doppler) excellent for short period after initialisation
	• DME fixing (ρ-ρ) better after time
	• VOR fixing (θ-θ) less accurate than DME but better than INS/Doppler after time

> • VOR/DME (ρ-θ) accuracy between the pure DME and pure VOR
> • Kalman filtering produces most accurate position from all inputs
>
> **Updating**
> • Kalman filter shows its calculated position, navigation uses this
> • Self-contained aids show their calculated position
> • When sure of position, pilot updates self-contained position to re-initialise
>
> **Airspace**
> • Basic (B-RNAV) fit required in parts of Europe, at and above FL 100
> • Required accuracy $\pm$ 5 nm (95% of the time)

Sample questions

1. What is the required accuracy (95% of the time) of a RNAV system fitted to an aircraft operating in airspace designated as B-RNAV airspace?

   a. $\pm$ 5° of track, $\pm$ 5% of distance flown
   b. $\pm$ 5% of distance flown
   c. $\pm$ 5 nm at all times
   d. $\pm$ 5° across track, $\pm$ 5 nm along track

2. Which of the following forms of fix is most likely to provide the most accurate input to a RNAV system?

   a. a DME range and a VOR radial from the same station
   b. 2 DME ranges from different stations
   c. 2 VOR radials from different stations
   d. 2 ADF bearings from different beacons

3. A RNAV system is initialised at 1240. Which of the following inputs is likely to be most accurate at 1245, if the aircraft travels at a groundspeed of 300 kts?

   a. An INS input
   b. A Doppler input
   c. Twin DME ranges
   d. Twin VOR radials

4. Look at the illustration below, which represents the CDI of an area navigation system. What does the indication tell the pilot?

a. He must turn left 3° to reach the next waypoint
b. He must turn right 3° to reach the next waypoint
c. He is 3 nm left of track
d. He is 3 nm right of track

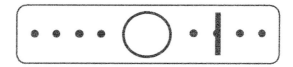

5. How often should the database of the FDSU be updated?

a. Every 90 days
b. Every 28 days
c. Every 14 days
d. After every flight

Chapter 17
Secondary Surveillance Radar

Introduction

The history and basic theory of secondary radar has been explained in Chapter 14. The original use of secondary radar, that of identification friend or foe, has been expanded, not only in the military context, but more importantly for civilian air traffic control purposes into Secondary Surveillance Radar (SSR). A transponder signal from an aircraft can now send any information required by ATC, without the crew making a radio call. That same information can be displayed either on the radar screen or on a separate screen alongside.

Because of the limitations of primary radar, several ATS units in the UK and abroad are now using only SSR to control aircraft within controlled airspace. This requires all aircraft which fly in that airspace to be equipped with SSR transponders capable of sending the information needed, but controllers are unable to see aircraft whose transponders are either unserviceable or not fitted.

Principle

A ground radar station, the 'interrogator', sends a rotating beam of pulse modulated signal in all directions in a form which the airborne transponder recognises as a request for information. The aircraft equipment decodes the interrogation, delays for a fixed short time, then sends back, 'transponds', an omnidirectional signal with the requested information on a different frequency. The ground receiver decodes the information, which may be combined with a primary response from a co-located primary radar on the primary PPI display, or be shown on a discrete display of its own.

Ground equipment

A typical SSR ground antenna is shown installed on a parabolic primary radar antenna in Figure 17.1. It consists of a rotating slotted array which transmits a narrow beam in the horizontal plane. It may be mounted, as

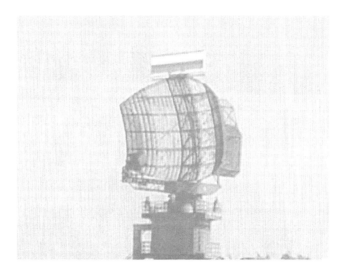

Figure 17.1   SSR ground antenna on PSR antenna.

shown, on top of a standard parabolic primary radar aerial, or be totally independent.

Frequencies

The interrogation signal is sent on 1030 MHz ($\pm$ 0.2 MHz). The transponded signal is returned on 1090 MHz ($\pm$ 3 MHz). For examination purposes, it may be convenient as an aide-mémoire to remember that 'the aircraft is higher than the ground'.

Current modes and codes

The signal from the interrogator is basically two pulses timed at specific intervals to request particular information. These pulse intervals are referred to as 'modes'. Mode 'A' sends two pulses at an interval of 8 microseconds (μs). These are referred to as the P_1 and P_3 pulses; the reason for the numbering will become clear later. The aircraft replies with another pair of pulses 20.3 μs apart, and between these two 'framing' pulses there is allocated space for 12 others. The presence or absence of each of these 12 pulses makes a specific pattern which is one of 4096 possible combinations. These combinations, allocated numbers for convenience as 'codes', provide identification of the target, which is the aim of the Mode 'A'.

Mode 'B' is available as a 'spare' for Mode 'A' transmissions. The pulse interval is 17 μs. However, it is not used, being superseded by the imminent arrival of Mode 'S'.

Mode 'C' is designed to provide altitude information to Air Traffic Control. A Mode 'C' transmission consists of two pulses with a 21 μs interval. This requests the airborne equipment to code the 12 information pulses of the response in accordance with the pressure altitude of the aircraft, which is digitally extracted from a pressure altimeter. Each combination of pulses represents 100 ft of pressure altitude, or one 'flight level'. To allow for small inaccuracies in different altimeter equipments, the pressure altitude shown on the controller's radar screen must be within ± 200 ft of the pilot's display.

Mode 'S'

Mode 'S' is intended to provide a datalink of information between aircraft and ATC, reducing the need for verbal communication. In order to allow slow transition between the systems, a Mode 'S' interrogator also sends interrogations and can interpret responses in Modes 'A' and 'C'.

In a Mode 'S' interrogation, the initial two pulses are followed by a long pulse consisting of a string of up to 112 bits (BInary digiTs). The bits are transmitted by making phase reversals in the long pulse. This string of bits forms a message, the first 24 bits of which is a unique address for the aircraft it wishes to interrogate. The combinations of these 24 bits provides over 16 million possible addresses, enough to include every aircraft registered in every state. The first digits in the address represent the region, the next few represent the State of Registration, and the remainder are allocated to the individual aircraft by the State of Registration.

This addressing system means that only the aircraft which is addressed will respond to the interrogation, removing the problems of 'fruiting', 'garbling' and 'over-interrogation' (see below). The transponding aircraft will reply with the information requested, coded in a similar phase modulated pulse. Either 112 bits or 56 bits may be sent, depending on what the interrogation has asked for.

The Mode 'S' normal transmission is this 'selective calling'. It can also make broadcast transmissions of information to all Mode 'S' aircraft, without needing a reply. However, if the interrogator does not know that a particular aircraft in its area is equipped with Mode 'S', it will never use its individual address to interrogate it. The controller will not know its details, although it will be responding to the Mode 'A' and 'C' interrogations which are sent out from the interrogating station simultaneously. The interrogator therefore sends out an occasional 'all call' message, which contains an extra pulse. A Mode 'S' transponder recognises this pulse as a request for every Mode 'S' equipped aircraft to respond with its address and basic information.

This basic information consists of the flight callsign, the capability of the transponder, and an altitude report in 25 ft intervals. The capability of the transponder is described in 'levels'. Level 1 has no datalink capability but recognises its individual address, effectively a Mode 'C' transponder with selective calling. Level 2 permits standard datalink communication between the aircraft and ground station in both directions, and is the minimum standard for future international flights. Levels 3, 4 and 5 increase the datalink capabilities beyond the standard flight information. This flight information includes magnetic heading, speed, roll angle, change in track angle, rate of climb or descent, and true track and groundspeed.

Mode 'S' has many advantages over the other modes:

- Increased capacity (999 tracks can be followed by the interrogator, as against 400 with Mode 'A')
- More accuracy in altitude reporting (25 ft intervals instead of 100 ft)
- Elimination of 'garbling', 'fruiting' and 'over-interrogation'
- Improved surveillance accuracy
- Possible use with future automated air traffic control systems
- Information can go both ways – can integrate with a flight management computer
- Datalink will relieve voice communication channels
- It forms a major part of an ACAS system (see Chapter 18)

Interrogation signal

Every radar transmission suffers from 'side lobes' which send energy out in directions other than that of the main beam. It would be possible for an aircraft to receive one of these side lobes and respond to that signal, interfering with other responses. These side lobes must be suppressed, and this is achieved by using an extra aerial which in a Mode 'A' or 'C' transmission sends an additional 'control' pulse $2\,\mu s$ after the first interrogation pulse. This control pulse is weaker than the interrogation pulse, in fact the control aerial polar diagram actually reduces the strength of the control pulse in the direction of the interrogation signal, forming a similar pattern to the limaçon of a VOR station. The target actually receives three pulses. If the second (the P_2 pulse) is weaker than the first (the P_1 pulse), it knows it is receiving the main interrogation beam, and responds. If the second is stronger, it knows it is outside the main beam, and does not respond. A representation of the two polar diagrams is at Figure 17.2. In a Mode 'S' interrogation, the control pulse is sent before the interrogator. This means that aircraft without Mode 'S' will not reply.

Aircraft aerial

An SSR antenna is used only for two very close frequencies in the UHF band. It is a small antenna, and often in the shape of a 'shark's fin', as in Figure 17.3,

(a)

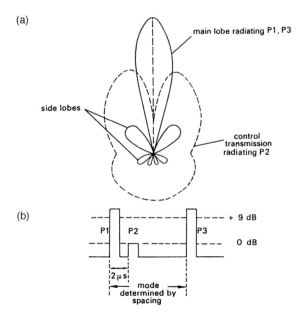

(b)

Figure 17.2 Side lobe suppression.

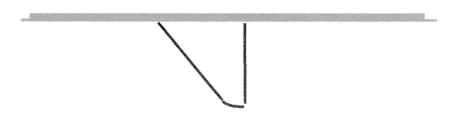

Figure 17.3 SSR aircraft antenna.

situated on the aircraft's underside. It receives and transmits omnidirectionally, so its polar diagram is a circle. It has been found that some signals may be blanked from this antenna by a wing when the aircraft is banked. Mode 'S' antennas are fitted on both top and bottom aircraft surfaces to reduce this.

Cockpit control unit

Most SSR equipment includes a control unit similar to that in Figure 17.4 below, which represents a unit transponding on Modes 'A' and 'C'. The code numbers are usually rotated by finger or thumb, but may be selected by push-buttons. The switch positions as shown have the following meanings.

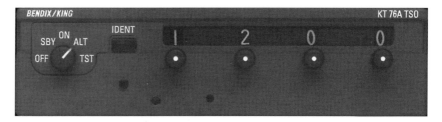

Figure 17.4 SSR control unit (courtesy Honeywell).

OFF   Power switched off

STBY   Standby – power available to the transponder, but no transmissions possible

ON   Equipment responds to Mode 'A' signals with the selected code

ALT   Equipment responds as above, and to Mode 'C' signals with the aircraft's flight level

TST   Equipment runs through a built-in test sequence to check itself

A light is provided to indicate that the equipment is replying to an interrogation. This light will also illuminate on completion of a successful TEST. There is also an IDENT or SPI (special position identification) button or switch. Selection of this sends an extra pulse 4.35 μs after the Mode 'A' code, and continues for approximately 20 s. This should be selected when ATC asks the pilot to 'squawk ident'. ATC will be able to identify the response by a double blip on the radar display.

Ground display

An SSR display usually shows a dot rather than the normal radar 'blip' at the position of the target, and weaker dots showing the position of previous returns. These dots give an indication of the aircraft's track and ground-speed. The current position will have a mark to emphasise it and designate the controller responsible for it, and beside that will be displayed a 'label' including either the Mode 'A' code, or the flight number or aircraft callsign as shown in Figure 17.5 below. Under that callsign is the Mode 'C' flight level, with an arrow if climbing or descending. Below that is the wake turbulence category, followed by the computed groundspeed. After that is space for other desired information to be displayed, such as the flight level to which the aircraft is cleared, or its destination. The label will flash if the crew select an emergency code.

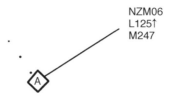

NZM06
L125↑
M247

Figure 17.5   SSR label.

Codes with specific meanings

Several Mode 'A' codes have specific meanings when selected by the pilot. The ICAO recognised codes are as follows.

7700   The aircraft is in a state of distress (Mayday)
7600   The aircraft cannot receive radio signals (radio failure)
7500   The aircraft is suffering from unlawful interference (hijack)
2000   The aircraft has had no code allocated by ATC. In the UK this is reserved for aircraft entering UK airspace from overseas.

There are also other standard codes used by ATC in the United Kingdom, some of which are listed below.

7000   (Conspicuity) The aircraft has no specific code allocated by ATC
7004   The aircraft is carrying out aerobatic manoeuvres
0000   The aircraft has a transponder malfunction (probably Mode 'C' failure)

Advantages of SSR

SSR has several advantages over primary radar.

(1)   Because a primary radar signal is reflected by the target, its maximum range depends on the fourth root of the power available in the signal. A SSR transmission only has to travel in one direction and be received by the aircraft, so the maximum range of the interrogation depends on the square root of the signal power. The maximum range of the transponded signal depends also on the square root of the power of the signal from the aircraft. This means that for a given range, a secondary radar requires less power, or for a given power output a secondary radar has greater range. Of course, the maximum range is governed by the UHF formula.
(2)   Because the signal received at the ground station is on a different frequency to that transmitted, there is no reflected ground return. In

fact, the only returns are those from transponding aircraft, so all false returns are avoided. The display is uncluttered unless it is combined with a primary radar.

(3)   Because the transponded message is coded, specific information requested by ATC can be received and displayed. A primary radar only indicates the target's position.

(4)   No manoeuvres are required from the aircraft to assist the control in identification – the SSR can provide all the identification needed.

(5)   No verbal communication is required other than to inform the aircraft of the correct code. Communications frequencies are left relatively quiet for other messages.

Disadvantages of SSR

The SSR system has the disadvantage compared with primary radar that the aircraft must be equipped with the correct transponder. Many small aircraft are not so equipped. There are also some problems to which Mode 'A' and Mode 'C' are subject, and, as seen earlier, Mode 'S' is designed to reduce these.

(1)   Because all equipments use the same frequencies, a transponder will reply to any interrogation signals, including those from interrogators whose operators are not in communication with the pilot. These replies may appear at different times and in different positions on the controller's screen. This is called 'fruiting', and can be reduced by allocating neighbouring radars different prfs and disregarding a sequence of returns at the wrong prf.

(2)   If two aircraft are in the interrogating beam at the same time, and close to each other so that their transponded signals may overlap at the receiver, it causes interference called 'garbling'. This will occur if the aircraft are less than about 1.7 nm apart. Manufacturers advertise 'killer circuits' in the receiver to separate the returns and reduce 'garbling'.

(3)   If an aircraft is being interrogated by several ground stations at once ('over-interrogation'), the transponder may suffer from a similar problem to the fruiting experienced at interrogators. The aircraft equipment may not be able to recognise the combination of individual signals as requests for information, and may not reply to one or more of them.

RT terminology

By tradition, and as a consequence of the military origins of transponders and their codes associated with that, the use of SSR is referred to by voice transmissions using the word 'squawk'. ATC will use the following expressions to request the pilot to make certain selections on the control unit.

SQUAWK xxxx (numbers)	Select the requested code numbers and Mode 'A'
SQUAWK CHARLIE	Select Mode 'C'
SQUAWK STANDBY	Select Standby (stop transmitting Modes 'A' and 'C')
SQUAWK IDENT	Operate the SPI switch
SQUAWK MAYDAY	Select 7700 and Mode 'A'

Automatic data surveillance

An alternative to the Mode 'S' transponder system which ICAO intends to bring into service, some manufacturers and countries have supported a different system using similar technology, but requiring no interrogation from the ground station. This 'automatic data surveillance' or ADS relies on the aircraft transmitting the information normally requested by Mode 'S', and other information also, as a broadcast to the ground station and to other aircraft.

The accuracy possible from satellite and other modern navigation systems means that the pilot of an aircraft, or at least his navigation computer, knows where it is in three dimensions at all times. A function of that navigation computer can automatically encode that position, and broadcast the data on a suitable frequency. Ground equipment can decode the information and plot it on a screen for a controller. The advantage is that the aircraft does not need to be in radar coverage, and potentially can provide information from the centre of oceans, using satellite communications.

Not only ground controllers find such an ADS broadcast (ADS-B) useful. Computers in other aircraft can receive the transmissions and compare them with their own. That comparison can be processed to provide collision avoiding action in the event of possible conflictions. The information would be displayed on a standard CDU.

Summary

Principle	Ground interrogates aircraft with 'mode' pulses, a/c replies with 'code' pulses.
Frequency	UHF – ground–air 1030 MHz, a/c–ground 1090 MHz. (Aircraft high)

Modes	• Interrogation P_1 and P_3 pulses Mode 'A' = 8 μs apart, Mode 'C' = 21 μs • P_2 for side lobe suppression, transmitted in limaçon pattern
Code reply	• 2 framing pulses 20.3 μs apart, 12 pulses on/off between = 4096 possible • Pulses in 4 groups of 3 for Mode A, each group gives number 0–7 • Extra SPI pulse 4.35 μs later
Range	VHF formula, 200 nm
Codes	• 7700 = Distress • 7600 = Radio failed • 7500 = Hijack
Mode 'C'	• Flight levels from a/c altimeter – limit $\pm$ 200 ft • Digitiser converts hundreds of feet to binary code • 11 pulses used – each bit shows 100 ft band
Controller	• STBY while changing • IDENT sends extra pulse for 20 s • 'Squawk' MODE 'A' for ATC • 'Squawk' MODE 'C' for pressure altitude
Mode 'S'	• 16 million binary codes • Can be used together with Modes 'A' and 'C' • Ground addresses individual aircraft • Information can go both ways – to and from FMS • Datalink will relieve voice comms channels • Major part of ACAS/TCAS
Fruiting	• Replies from other interrogators • Reduced by allocating adjacent stations different prfs
Garbling	• Replies from aircraft close together interfere • Removed by 'killer circuits'
Built-in test	Press button – tests internal circuits if OK, light on for 15 s

Sample questions

1. What SSR code would you select on Mode 'A' if you were entering UK airspace from overseas, and had not been allocated a specific code?

   a. Standby
   b. 2000
   c. 7000
   d. 0000

2. The P_2 signal of an SSR interrogation:

   a. Is stronger than the Mode 'A' pulses?
   b. Comes after the Mode 'A' pulses?
   c. Is stronger than the Mode 'A' side lobes?
   d. Comes after the Mode 'A' side lobes?

3. Which of the following is untrue?

   a. Mode 'S' SSR addresses individual aircraft
   b. Mode 'S' SSR allows more accurate height readout than Mode 'C'
   c. Mode 'S' SSR uses twice as many codes as Mode 'A' or Mode 'C'
   d. Mode 'S' SSR can be used together with Mode 'A' and Mode 'C'

4. Which of the following is not an advantage of secondary radar over primary radar?

   a. Ground returns are eliminated
   b. Every aircraft within range paints on the screen
   c. The target aircraft identifies itself
   d. Power required by the transmitter is reduced considerably

5. Your aircraft is being hijacked. What SSR code should you select on Mode 'A'?

   a. 7700
   b. 7600
   c. 7500
   d. 7007

6. Why are SSR transmissions in pairs of pulses?

   a. To prevent random noise being falsely identified
   b. To provide side lobe suppression
   c. To indicate the mode of interrogation
   d. To avoid dividing the response time by 2 to find the range

7. How many codes are available in Mode 'A' of SSR?

   a. 3
   b. 7777
   c. 4069
   d. 4096

8. What does the 'SPI' facility of an SSR do?

   a. It sends an extra pulse for 20 s
   b. It switches the Mode 'A' response on and off for 20 s
   c. It sends an extra pulse for 1 min
   d. It switches the Mode 'A' response on and off for 1 min

9. Mode 'C' gives indications to the ground about the aircraft's altitude. If the indications are more than a certain amount in error, it must be switched off. What are the limits?

   a. $\pm$ 50 ft
   b. $\pm$ 100 ft
   c. $\pm$ 150 ft
   d. $\pm$ 200 ft

Chapter 18
Collision Warning Systems

Introduction

It became obvious that if secondary surveillance radar could be used to provide information to a ground station about the aircraft carrying a transponder, then it might be possible to provide the same information to an airborne station. The vast increase in computer power over the years since SSR came into general use has given that idea reality. It has also allowed systems to be developed which can process the information received from these transponders and give directions to an aircraft fitted with one of them to prevent it colliding with the transponding aircraft.

The generic term for all these systems is 'Airborne Collision Avoidance System' or ACAS. This includes the systems developed in the United States called 'Traffic Alert and Collision Avoidance System' or TCAS. TCAS is at the time of writing the only system which provides the requirements of ACAS. ICAO defines the requirements for ACAS in Annex 10, and that term will be used throughout for their required equipment. Examination questions can be expected to follow the same principle. References to TCAS apply to available equipment, references to ACAS apply to the requirements.

Principle of operation

Basically, the airborne equipment consists of an interrogator (transmitter) and receiver on SSR frequencies (1030 MHz and 1090 MHz respectively). The transmitter need not have great power, because a pilot is unlikely to be interested in traffic at long ranges. For the same reason, it requires a short prf and can sweep quite quickly. The system sends an SSR Mode 'S' interrogation (which includes interrogations in Modes 'A' and 'C') and receives the code from the other aircraft. It uses directional antennas to determine the relative direction of the transponded signal, and calculates the range from the time between transmission and reception (less the normal transponder fixed delay).

ACAS I

The equipment sends out interrogation pulses and awaits replies. When a signal from a Mode 'A' transponder is received, the direction can be determined and range calculated. The information on range and bearing can be displayed on a screen for the pilot, and if the range is below a certain 'trigger distance' an audible and/or visual warning can alert him to the presence, direction and range of the other aircraft. He can then look in that direction and take any avoiding action he deems necessary.

The information from a Mode 'C' transponder can be decoded, and the transponding aircraft's relative altitude displayed on an instrument. Alerts need not be given about traffic which may be conflicting in plan but have adequate vertical separation, although the display will normally still show its position. A system which can display this information and give 'traffic advisories (TAs)' like this is called a TCAS I or ACAS I system. ACAS I gives approximately 40 seconds warning of a possible collision.

Despite being superseded by ACAS II for international air transport operations, ACAS I equipment is simpler and cheaper to install, and provides TAs which a pilot can use to direct his eyes towards a possible threat. This has been realised as a distinct benefit to aircraft operating under Visual Flight Rules. Helicopters especially have a technical difficulty in using TCAS II equipment, and lightweight ACAS I systems have been developed to warn pilots of approaching aircraft. Figure 18.1 shows the monochrome display of a system which satisfies the requirements of ACAS I.

ACAS systems are an extension of the SSR transponder. They have a sophisticated computer processor which tracks the relative position of the

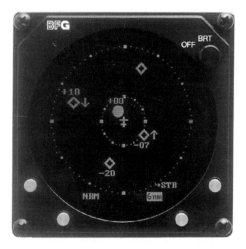

Figure 18.1 Monochrome traffic alerting display (courtesy BFGoodrich Avionics).

transponding aircraft in three dimensions. While tracking the transponding aircraft it calculates not only the range and relative altitude, but also the rate of change of both. The processor can then calculate the time to, and the separation at, the closest point between the two aircraft. If the closest point is likely to be within a certain distance, it generates a TA to advise the crew of an aircraft of that fact. It will ignore signals from aircraft outside 10 000 ft of its own altitude.

If the 'intruder' continues to close, the TA symbol changes once the intruder is within 6 nm horizontally and 1200 ft vertically. This traffic is defined as 'proximate', although there may be no direct threat of collision.

ACAS II

ACAS II requires the system to provide avoiding action if appropriate. If the computer calculates that the approaching traffic is a potential collision threat, it then calculates the best method of avoiding the confliction, an 'escape manoeuvre'. Because the directional antennas are not particularly accurate, the only safe escape manoeuvres are in the vertical plane, so the directions are to climb or descend at nominated rates. These directions are referred to as 'Resolution Advisories' or 'RAs'.

A RA will give quite exact directions to avoid the collision. A climb or descent if advised should be entered and continued at set rates, as indicated below. The advisory information will continue, and the advice will change during the manoeuvre, until the confliction is past.

If two aircraft equipped with ACAS II appear to be conflicting with each other, their Mode 'S' datalinks can 'talk' to each other, passing data in their Mode 'S' transmissions. This will co-ordinate their RAs to optimise their separation and minimise disruption. ACAS II equipment can track up to 45 aircraft, and display up to 30 of them. It can co-ordinate RAs to avoid up to 3 conflicting aircraft.

If an aircraft's ACAS II equipment generates a RA, its Mode 'S' transponder can inform the ground interrogator of the manoeuvre automatically. Otherwise the crew must inform ATC of any deviation from the flight path which ATC have allocated it. Pilots are expected to make altitude changes to follow RAs, provided they tell ATC immediately, using standard phraseology.

Below 1000 ft above ground level, the ACAS II system will not generate any RAs. Nor will an ACAS aircraft be given a RA to descend below 1000 ft agl. Any intruder which has an apparent vertical speed above 10 000 ft/min up or down will not generate a RA.

ACAS III

ICAO has issued more advanced specifications for a future system to be called ACAS III. This will use the Mode 'S' transponded signal from the other aircraft, with the intention of providing horizontal RAs. However, at the time of writing that lies in the future.

ACAS II details

Air traffic control requirements

The European Civil Aviation Conference (ECAC) states decided that by 1 January 2000, every fixed wing aircraft with turbine engines and more than 30 seats in their area must carry a serviceable TCAS II system with software version 7. The same applies to any aircraft with a maximum certificated take-off mass greater than 15 000 kg. In addition, all UK registered aircraft in those mass or capacity brackets must carry it wherever they are in the world. Although the commercially available TCAS II system was intended to fulfil ACAS II requirements, version 7 was the first to comply fully with the ICAO ACAS II requirements. Because of equipment availability problems, the JAA allowed certain temporary exemptions. However, the requirement will be extended to include all aircraft over 5700 kg or with more than 19 passenger seats from 1 January 2005.

ACAS II functions

According to ICAO Annex 10, ACAS II is designed to have the following functions:

- surveillance and generation of traffic advisories
- detection of collision threats and generation of resolution advisories to avoid them
- co-ordination with other suitably equipped aircraft
- communication with ground stations

ACAS II inputs

To function properly, ACAS II requires the following inputs:

- aircraft address code
- air–air Mode 'S' transmissions received by the Mode 'S' transponder
- own aircraft's maximum cruising true airspeed capability
- pressure altitude
- radio altitude (see Chapter 20)

Intruder's requirements

The main problem with ACAS is that it can only detect and alert to aircraft which have operating transponders. It can only deconflict with aircraft which have operating and serviceable altitude reporting (Mode 'C') transponders. It is therefore important that all aircraft, even those not receiving a radar service, have their transponders switched on in both Mode 'A' and Mode 'C'.

Airborne equipment

Antennas

ACAS II uses two antennas, one above and one below the aircraft. The top antenna is a direction finding one, as is often the bottom one. Several installations may use an omnidirectional receiver antenna on the bottom surface, but no directional information is then available if the transponded signal is not received by the top antenna. These antennas are kept separate from the normal SSR antenna.

Control unit

A typical TCAS control unit is similar to an SSR controller, with the additional functions of selecting either Traffic Alert or both Traffic Alert and Resolution Advisory functions. A TCAS/Mode 'S' transponder control unit is shown below, in Figure 18.2. The VFR function allows an automatic change in transponder code to a preselected VFR code (7000 in the UK).

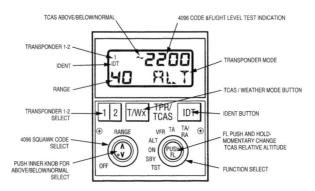

Figure 18.2   TCAS control unit (courtesy Honeywell).

Displays

Symbology

There are four different symbols which may appear on the TCAS II cockpit display. These appear in a position on the display relating to the relative position of the other aircraft, although if the system cannot determine the bearing of that aircraft it will show the relevant symbol in a convenient position with the words 'no bearing'. Manufacturers claim a bearing accuracy of 2°.

The symbol displayed is selected by the equipment depending on the intruder's position and closing rate. A TA which is not a threat will appear initially as an open white (or sometimes blue) diamond on the display. This means it is more than 6 nm or more than 1200 ft vertically (if a Mode 'C' signal is received) away from your aircraft. A solid diamond indicates that the other traffic is within that safe zone (i.e. 'proximate'), but the computer calculates it is still not a threat.

If the computer calculates that the intruder is potentially hazardous, the symbol will change to a solid yellow circle. A voice TA will be given of 'Traffic Traffic'. About 10–15 seconds later, if the intruder is assessed as an actual collision threat, the symbol will change to a red square, and a voice command will give an indication of the computed RA, which will be displayed exactly on the cockpit display.

If the intruder is transponding with Mode 'C', the symbol will have beside it the relative altitude in hundreds of feet. If the relative altitude is changing by more than 500 ft/min a plus or minus sign will also appear. Examples of the symbols are given in Figure 18.3 below.

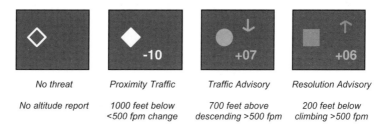

| No threat | Proximity Traffic | Traffic Advisory | Resolution Advisory |
| No altitude report | 1000 feet below <500 fpm change | 700 feet above descending >500 fpm | 200 feet below climbing >500 fpm |

Figure 18.3 TCAS II symbols.

RA VSI

The stand-alone cockpit display is on a modified vertical speed indicator, an example of which is shown in Figure 18.4 below. The basis of the instrument is a VSI as described in Volume 3 of this series. Under the centre of the instrument is a representation of the ACAS aircraft with a ring of dots at a

Figure 18.4 RA VSI 'descend descend' (courtesy Honeywell).

selected range. The relevant symbol for the 'intruder' is shown in the correct horizontal position in relation to the aircraft symbol and the range ring. When a RA is computed, an outer ring shows two colours, red and green, to indicate the safe and unsafe rates of climb or descent to the pilot.

In Figure 18.4, the ACAS equipment is fitted to the aircraft on the left. The short, fat, green arc between 1500 and 2000 ft/min descent is slightly wider than the much longer red band, which shows the large unsafe area. As the pilot follows the directions, the green arc may move around the rim of the instrument. Other versions use rows of red and green lights to display the same information. In some cases the safe option is to neither climb nor descend, in which case the green arc will appear around the zero mark.

Radar display

On a radar display, such as an AWR screen, the 'intruder' symbol is shown in the position on the screen corresponding to its position in relation to the ACAS aircraft. The symbol is made more prominent than any other information. The same applies to the horizontal situation screen of an EFIS display, which is described in *Navigation* in this series. As with other EFIS inputs, the display shows a map of the horizontal situation in relation to the aircraft, and the TCAS symbols appear in their relative positions.

EFIS vertical commands

If there is a radar screen in the cockpit, the RA VSI may not show the horizontal position of the other aircraft. It will appear as a simple VSI with the addition of the green and red lights or bands to indicate the safe and unsafe rates of climb or descent. On an Attitude Director Indicator (ADI) the unsafe and safe bands may be shown as red/green bands on the VSI marked on the side of the AI. They may also be included in the directing marks on the face of the instruments, usually in red. An example of both expanded HSI and ADI parts of an EFIS display is shown in Figure 18.5 below, indicating a climb command. There is no VSI, but the red 'tram lines' on the ADI display indicate a 5° climb angle required for the initial avoidance. Descent RAs will not be given if the aircraft is below 1000 ft agl, which is why the radio altimeter input is required to the unit.

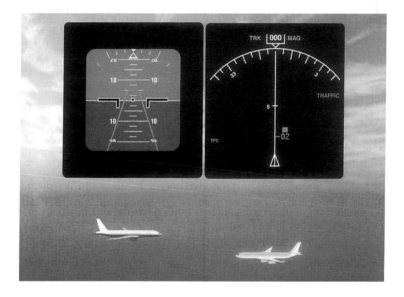

Figure 18.5   EFIS RA 'climb climb' (courtesy Honeywell).

Aural advisory messages

TCAS II has a vocabulary of phrases which it uses to pass its advisory commands to the pilot. These are listed in Table 18.1 with the actions which the pilot must carry out to avoid the confliction.

Table 18.1 TCAS messages

Voice Command	Performance Required
Traffic, Traffic	Identify traffic and prepare for possible RA
Maintain Vertical Speed, Maintain	Maintain existing rate of climb or descent
Climb, Climb	Climb at 1500 ft/min
Descend, Descend	Descent at 1500 ft/min
Increase Climb	Climb at 2500 ft/min
Increase Descent	Descend at 2500 ft/min
Climb, Climb NOW	Initiate change from descent into climb
Descend, Descend NOW	Initiate change from climb to descent
Climb, Crossing Climb	Climb at 1500 ft/min through the intruder's altitude
Descend, Crossing Descend	Descend at 1500 ft/min through the intruder's altitude
Adjust Vertical Speed, Adjust	Reduce climb or descent to the rate indicated
Monitor Vertical Speed	No manoeuvre required, keep out of the red area
Maintain Vertical Speed, Crossing Maintain	Maintain rate of climb or descent while passing through the intruder's altitude
Clear of Conflict	Resume normal flight, return to cleared altitude if under ATC control

Pilot's actions

The system is designed for the aircraft to follow the exact avoiding action within 5 seconds of an initial command and 2–3 seconds for subsequent commands. The required manoeuvres are not meant to be violent, a maximum of 0.25 'g' is intended. If the commands are followed exactly, including the 'softening' commands issued to reduce rates of climb or descent after the initial command, the system is designed to restrict altitude changes to a maximum of 500 ft.

If an ACAS RA is given when an aircraft is under ATC control, the pilot is required to follow the command and inform ATC 'ACAS climb' or 'ACAS descent' as appropriate. Once the danger is past, on receipt of the message 'clear of conflict', he must return to his cleared level as soon as practicable.

Monitoring system

A system of monitoring both the ACAS transmission power and the necessary inputs is provided. This monitoring also covers any failure in the equipment which would adversely affect its capacity to provide TAs or RAs. If the monitoring system detects a problem, it switches off the interrogator, warns the crew, and sends a Mode 'S' message to ground interrogators.

ACAS II errors

It is possible, but progressively more unlikely, for false RAs to be generated by other transmissions on the transponder frequency. These 'parrots' from ships or parked aircraft (those without the Mode 'S' facility of switching the transponder off on the ground) can be minimised by switching the system to a 'TA only' mode at low level. The same action can avoid RAs being given from aircraft on parallel runway approaches.

Light sensitive warning system

Realising the benefits of collision warning systems, but attempting to reduce cost for light aircraft, a system was developed to detect not a radio signal, but high intensity lights, such as are carried by many aircraft as anti-collision beacons. The strobe warning systems has light detectors, similar to solar electricity generating panels, arranged to give coverage around the planform of the aircraft, usually one on each wingtip and one on the tail. The simple display consists of lights at compass points around a representation of the aircraft, which illuminate to indicate the approximate direction of the light which has been detected. No indication of distance can be given, and the range depends on atmospheric conditions, such as dust haze and water droplets in cloud or precipitation.

ACAS summary

Principle	• Aircraft carries transponder AND interrogator – to/from other aircraft • Gives/receives position and flight level if available • Computer assesses position, threat and necessary action – displays it
Nomenclature	ACAS is ICAO requirement, TCAS is US or commercial name
Range	Replies from aircraft up to 30 nm away

Warnings	• TA = Traffic Advisory = no assessed threat at present • RA = Resolution Advisory = threat, action given
ACAS I	Simple – • designed to aid visual acquisition 　　　　• Displays range and approximate bearing, relative height (Mode 'C' or 'S') 　　　　• Warns (TA visual/audio) 40 s from potential collision
ACAS II	• More accurate, offers RAs in vertical sense as well as TAs • If both with Mode 'S', computers talk, co-ordinate RA • Uses directional aerials, transponder, processor and CDU • Power reduces in areas of high traffic density
RA VSI	Used for TCAS/ACAS II, red arc shows RoC/D to avoid, green shows safe
HSI display	Horizontal display of all transponding traffic • AWR or EFIS, or separate TCAS display • No threat = hollow white or blue diamond • 'Proximate' (within 6 nm/1200 ft) = solid white or blue diamond • TA = yellow circle • RA = red square • RA VSI may be added on VSI strip of EFIS
ACAS III	Uses Mode 'S' to give horizontal advisories also.

Sample questions

1. ACAS II gives information to the pilot about collision threats. What is the advice it provides?

   a. It tells the pilot the range of a possible collision threat, but no bearing
   b. It tells the pilot the bearing of a possible collision threat, but no range
   c. It tells the pilot range and bearing of a possible collision threat
   d. It advises the pilot of avoiding action in the vertical plane

2. What is the normal range of an ACAS installation?

   a. 120 nm
   b. 60 nm
   c. 30 nm
   d. 10 nm

3. What information is available from an ACAS I installation?

    a. Range and direction of conflicting traffic
    b. Range, direction and track of conflicting traffic
    c. Direction of conflicting traffic only
    d. Range of conflicting traffic only

4. What indication should a pilot expect on a TCAS I installation if the equipment assesses that another aircraft is a probable collision threat?

    a. Resolution advisories are given on a plan display only
    b. Audio and/or visual signals give traffic advisories only
    c. Audio and/or visual signals in plan give traffic or resolution advisories
    d. Resolution advisories may be given in the vertical plane

5. How much warning will a TCAS I give a pilot before a possible collision?

    a. 10 s
    b. 20 s
    c. 30 s
    d. 40 s

6. Which of the following is true about ACAS II compared with ACAS I?

    a. Traffic advisories are given in the vertical plane as well as horizontal
    b. Resolution advisories are given in the vertical plane
    c. Resolution advisories are given in the horizontal plane
    d. Resolution advisories are co-ordinated between all conflicting aircraft

7. What indication would a pilot see on his ACAS II display if an aircraft was assessed as requiring resolution advisory action?

    a. A white square
    b. A blue diamond
    c. A yellow circle
    d. A red square

8. If a pilot sees a solid white diamond on his EFIS, what does that tell him?

    a. This is a traffic advisory which will pass well away from his aircraft
    b. This is a traffic advisory which will pass close to his aircraft
    c. This is a resolution advisory which the pilot must turn to avoid
    d. This is a resolution advisory which the pilot must change height to avoid

9. What is needed for an aircraft's ACAS I equipment to indicate the relative height of a target which is a possible collision risk?

   a. Aircraft and target must have ACAS I equipment
   b. Aircraft must have ACAS I equipment, the target must have ACAS II
   c. Only the aircraft needs ACAS I, the target must have a serviceable Mode 'C'
   d. Only the aircraft needs ACAS I, the target must have a serviceable Mode 'A'

10. What is needed for an aircraft's ACAS equipment to give horizontal avoiding action from target which is a probable collision risk?

    a. Aircraft and target must have ACAS III equipment
    b. Only the aircraft needs ACAS III, the target must have Mode 'S'
    c. Only the target needs ACAS III, the aircraft can have ACAS II
    d. Only the aircraft needs ACAS III, the target can have ACAS II

Chapter 19
Microwave Landing System

Introduction

An ILS installation suffers from major problems in avoiding multipath errors from its site position, as seen in Chapter 8. An ILS installation on many aerodromes cannot be sited free enough of surrounding terrain or buildings to allow even Cat I approaches, let alone the Cat III approaches they desire. There are also other problems, which were recognised in the late 1960s. Among these were the limited number of channels available, and the restriction on approach paths which force aircraft with different performances to queue up to use it. There were also problems with ILS signals in precipitation, especially snow.

In 1972 ICAO published an operational requirement for a new type of non-visual approach and landing guidance system. This was to use a method called 'Time Referenced Scanning Beam (TRSB)' in the SHF band. Signals at these frequencies are commonly called 'microwaves', so the system became known as the microwave landing system or MLS.

Time referenced scanning beams

Basic principle

The idea is for a ground station to sweep a narrow fan-shaped beam at a very accurate constant speed from one side of a sector to the other, then back again after a specific time interval (the 'guard time'). The signal will be received twice at the airborne equipment, and the time between each signal (less the guard time) relates to the angle from the reference line, which is the position of the beam when it starts its sweep. This can be seen in Figure 19.1 below.

One fan-shaped beam sweeps horizontally to provide a position line in azimuth, as in Figure 19.1. At a different time, a horizontally orientated fan sweeps up and down in a similar fashion to give a position line in elevation. The angle of approach is now known in both azimuth and elevation, and can be displayed in a similar fashion to that of ILS. The third part of the system consists of an accurate DME (precision DME or DME/P) signal to show the

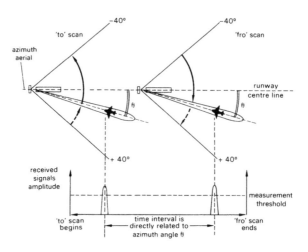

Figure 19.1 TRSB beam principle.

aircraft's position in range from the station. The aircraft's position can thus be determined in three dimensions.

Multiplexing

Technology allows every piece of information from each of the azimuth and elevation beams to be obtained from signals on the same frequency. Each piece of information requires a very short time to obtain it. After one piece has been received it is used and stored until it is replaced. Meanwhile, another piece of information can be received, and again used and stored; then another. The total time taken to receive every piece of information required for the MLS system to function in this fashion is about 84 milliseconds. This is divided into specific periods or bands in which the individual pieces of information are transmitted (and received), as shown in Figure 19.2. This is called multiplexing.

In addition to guidance information, auxiliary information is also sent during the multiplex transmission. this includes the station identification, safety information such as the minimum safe glideslope angle, and more sophisticated information such as system condition, weather and runway conditions which can be displayed on modern cockpit displays if fitted. Every piece of information includes a 'preamble' to synchronise and prepare the relevant part of the airborne equipment for the following 'function transmission' which contains the actual information signal, for example the beam sweep.

Beams can also scan in the opposite direction, away from the approach path, to provide guidance to aircraft in the missed approach segment. These are also useful on climb out after take-off. There are also test pulses, to check

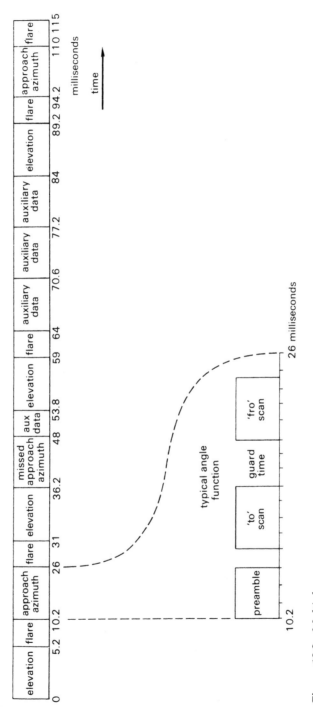

Figure 19.2   Multiplex sequence.

the serviceability of the system, and indicator pulses to give general guidance in the area between the approach and missed approach segments to guide the aircraft into the approach segment. Other signals may be transmitted to give guidance at the flare on touchdown for Category III approaches.

The time is not equally divided. Three elevation signals are received for every azimuth signal. This indicates the greater danger of a rapid change in elevation angle compared with a change in azimuth angle. There are 40.5 elevation scans every second, and 13.5 azimuth scans.

Frequencies

There are 200 allocated channels, spaced 300 kHz apart in the band between 5031.00 and 5090.70 MHz. Each station uses one channel for all its transmissions except the DME/P, which uses similar frequencies to a normal DME. The DME/P frequencies are automatically selected.

Azimuth coverage

In the approach segment, the horizontal area scanned by the guidance beams is 40° either side of the centreline, out to 22.5 nm from the station, although proportional guidance may be restricted to within 10° of the centreline. Vertically, the beams give guidance between 0.9° and 20° above the horizontal, up to 20 000 ft, although again proportional guidance may be restricted to a maximum elevation of 7.5°. There is a region over the runway in which coverage is provided from 8 ft up to 2000 ft and out to 150 m either side of the centreline. The missed approach segment is 20° either side of the centreline, and from 0.9° to 15° vertically, out to 10 nm from the station and up to 10 000 ft. A horizontal representation of the azimuth guidance coverage is at Figure 19.3, and a vertical representation at Figure 19.4.

Elevation guidance coverage

The specification for elevation coverage is slightly different. It only covers the approach sector, out to at least 20 nm, and within a horizontal angle which at least corresponds to that within which the azimuth proportional guidance is available. In elevation, the minimum coverage is from 0.9° to 7.5° above the horizontal, although it is recommended that it cover the whole of the azimuth approach sector. Representations of the coverage are at Figures 19.5 and 19.6 below.

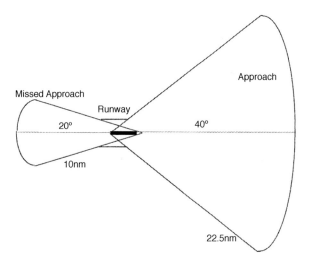

Figure 19.3   MLS azimuth guidance horizontal coverage.

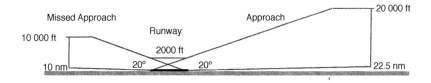

Figure 19.4   MLS azimuth guidance vertical coverage.

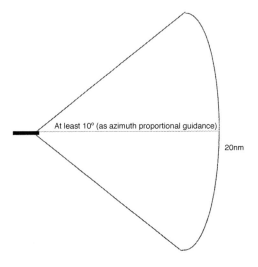

Figure 19.5   MLS elevation beam horizontal coverage.

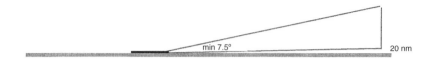

Figure 19.6   MLS elevation beam vertical coverage.

DME/P

Like the DME stations used for ILS ranges, the DME/P is electronically adjusted to give ranges from touchdown. Correct indications are available within the coverage of the guidance beams in the approach sector.

Use of MLS

Approaches

The aim of the system is to allow aircraft to make approaches from whichever direction they may be coming towards the runway, with guidance down an accurate path to a decision height or to a position from where an automatic landing can be made. The circling approach requires a computer, or central processing unit (CPU) which can compute a safe three-dimensional approach path and compare the aircraft's actual path with the pre-computed one.

Aircraft without such a sophisticated CPU can still use the equipment, but only on the 'segmented' approaches. A zero-reader type display, as described in the Chapter 8, can direct such a segmented approach. If the DME/P is not available, approaches can still be made, but only on a straight, ILS type, approach.

Accuracy

ICAO Annex 10 lays down the accuracy which must be achieved by a MLS system. Every requirement must be met at least 95% of the time. The most important requirement must be met at the MLS approach reference point, which like the ILS reference point is 50 ft above the point where the runway centreline intersects the threshold. Azimuth accuracy must be within 20 ft, elevation accuracy within 2 ft.

Monitoring

Each ground station monitors its output, and switches off the signal if any of the following takes place.

- Transmitter power reduces below minimums
- An error in the preamble transmissions
- If the permissible accuracy at the reference point is exceeded for more than one second
- The time synchronisation of the signals fails

Errors

MLS is not immune from interference from vehicles or aircraft near the transmitter. Multipath errors are still present. However, the interference from nearby fixed objects can be reduced by interrupting the scanning beam as it passes through such obstacles. At long ranges, the signal will have bent slightly so that the interruption is not noticed, and at short ranges the aircraft will not be in that part of the beam.

Airborne equipment

Despite being mandated by ICAO, MLS has not been universally accepted as an ILS replacement. Especially in the United States, DGPS has been put forward as an alternative, and MLS development was discouraged. Such uncertainty has led to manufacturers producing receivers which can accept signals from both new systems and display them. In fact, in order to minimise weight and space requirements during the transition from ILS to the new system, the receivers include the facility to receive and display ILS signals also. These are 'multi-mode' receivers.

Computer and microengineering technology allows inputs from each of the systems' aerials to be individually processed in modules within the same 'box' fitted to the aircraft. The various modules can then be selected for further processing within a main processor, and displayed on either a CDI, HSI or EFIS. Monitoring of the various signals can be carried out within the different modules, and failure indications displayed as necessary.

Advantages of MLS over ILS

- The primary advantage is the freedom from siting errors. Large objects do not create multipath errors
- MLS allows a variety of approaches, from many directions, including curved approaches, and a variety of possible glideslopes to cater for each aircraft's ideal performance
- Aircraft can approach from different directions, rather than having to fly

to a position on the extended centreline of the runway in use. This
expedites traffic flow in terminal areas
- The different possible approach directions allow aircraft to effectively
  overtake as they descend, increasing the number of aircraft which can
  approach at one time and increasing the capacity
- Guidance is more accurate with MLS than ILS

Summary

Principle	• Time referenced scanning beam; time between signals = angle • Large but narrow fan beam, delay at mid-scan before return • Precision DME gives range confirmation in same signal • Pilot selects desired path, indication gives guidance to that
Frequency	SHF band, approximately 5 GHz, 200 channels, each system one channel
Cover	Generally same in azimuth and elevation, allows pilot selected approach
Azimuth	• Approach $\pm 40°$ from centreline, up to 20° and 20 000 ft out to 22.5 nm • Beam scans 13.5 times/s. Missed approach up to 10 000 ft, out to 10 nm
Elevation	$+0.9°$ up to 7.5° minimum (20° ideal) up to 20 000 ft. Beam scans 40.5 times/s
Advantages	• No siting errors • Allows varied approaches, curved in azimuth, desired glideslopes • Expedites flow in terminal areas • Large capacity • Accurate guidance
Accuracy	20 ft horizontally, 2 ft vertically at MLS approach reference point
Errors	Interference from vehicles or aircraft near transmitter

Sample questions

1. Up to what angular distance from the centreline would a pilot receive correct MLS azimuth indications?

   a. 20° either side
   b. 30° either side
   c. 40° either side
   d. 60° either side

2. What frequency bands does MLS use?

   a. UHF for azimuth and elevation
   b. UHF for azimuth, SHF for elevation
   c. SHF for azimuth, UHF for elevation
   d. SHF for azimuth and elevation

3. An MLS azimuth guidance signal may be received:

   a. Within 20° of the centreline, from 1 to 10° above the ground?
   b. Within 20° of the centreline, from 1 to 20° above the ground?
   c. Within 40° of the centreline, from 1 to 10° above the ground?
   d. Within 40° of the centreline, from 1 to 20° above the ground?

4. What is the principle of operation of an MLS receiver?

   a. Comparing the strengths of 2 signals in each plane
   b. Comparing the phases of 2 signals in each plane
   c. Comparing the time between 2 signals in each plane
   d. Comparing the frequencies of 2 signals in each plane

5. Which frequency would be most appropriate for an MLS station?

   a. 50.59 MHz
   b. 505.9 MHz
   c. 5059 MHz
   d. 50.59 GHz

6. Which wavelength would be most appropriate for an MLS station?

   a. 6 mm
   b. 60 mm
   c. 60 cm
   d. 6 m

7. Which of the following is not an advantage of MLS over ILS?

   a. It expedites flow in terminal areas
   b. It has a large capacity
   c. Guidance is more accurate
   d. Warning is given if the signal fails

8. Which of the following is not an advantage of MLS over ILS?

   a. It allows curved approaches
   b. It has no siting errors
   c. It allows varied glideslope angles
   d. The beam does not suffer interference problems

9. What is the cover of the glidepath beam of an MLS station?

   a. $\pm 20°$ from the centreline, out to 20 nm, $1°$ to $20°$ in elevation up to 20 000 ft
   b. $\pm 10°$ from the centreline, out to 10 nm, $1°$ to $20°$ in elevation up to 20 000 ft
   c. $\pm 20°$ from the centreline, out to 10 nm, $1°$ to $20°$ in elevation up to 10 000 ft
   d. $\pm 40°$ from the centreline, out to 20 nm, $1°$ to $20°$ in elevation up to 20 000 ft

10. What is the cover of the azimuth beam of an MLS station?

   a. $\pm 20°$ from the centreline, out to 20 nm, $1°$ to $20°$ in elevation up to 20 000 ft
   b. $\pm 40°$ from the centreline, out to 20 nm, $1°$ to $20°$ in elevation up to 20 000 ft
   c. $\pm 20°$ from the centreline, out to 40 nm, $1°$ to $20°$ in elevation up to 40 000 ft
   d. $\pm 40°$ from the centreline, out to 40 nm, $1°$ to $20°$ in elevation up to 20 000 ft

11. If the frequency of the horizontal beam of an MLS station is 5.04 GHz, which would be the frequency of the vertical beam?

   a. 2.52 GHz
   b. 5.04 GHz
   c. 10.08 GHz
   d. 50.40 GHz

12. How many times does the MLS beam scan horizontally compared with vertically?

    a. Three times vertically to once horizontally

    b. Three times horizontally to once vertically

    c. Once vertically to once horizontally

    d. Twice horizontally to once vertically

13. Which of the following is an error of MLS?

    a. Interference from vehicles or aircraft close to the transmitter

    b. Siting error

    c. False glidepath at approximately twice the angle of the correct one

    d. Scalloping

14. How is range measured by the MLS?

    a. DME ranges are found from normal installations

    b. A separate precision DME gives ranges from the centre of the airfield

    c. A precision DME gives ranges from touchdown

    d. Azimuth and elevation signals combine to give a 3-dimensional picture

Chapter 20
Radio Altimeters and Altitude Warnings

Introduction

A barometric altimeter cannot measure the height of an aircraft above the terrain. Even with the local pressure setting (QFE) as a datum it will only measure the height above that datum. A continuous measure of height above terrain can be obtained from an altimeter which sends radio signals to the ground and receives the reflection, because the time for the waves to reach the ground and return to the aircraft can be measured and the range (in this case the height) calculated.

Some radar altimeters were developed which sent pulses to the earth and measured the time delay until reception. The time delay, multiplied by the speed of radio waves C and divided by two gave the height to a reasonable degree of accuracy. However, modern equipments use a different technique, a frequency modulated continuous wave (FMCW) signal which provide greater accuracy at low heights, which is where ground clearance is most important.

Principle of operation

The radio altimeter (sometimes referred to as a radalt) uses a technique similar to the Doppler principle. However, instead of sending a fixed frequency and measuring the Doppler shift to obtain the relative speed, it sends a continuously, regularly, changing frequency and measures the time taken between transmission and reception of each frequency.

The carrier frequency, in the SHF band at around 4.3 GHz, is frequency modulated by a further signal of about 300 Hz to produce varying frequencies between about 4200 MHz and 4400 MHz. The reflected signal, having travelled to the ground and back, is received at the instrument at a frequency removed from the one being transmitted at the time. Over a short period of time, when the height is unchanged, although the received and transmitted frequencies will change, the difference in frequency between the transmitted and received frequencies will remain constant. That difference is a function of the height above ground.

Indicators

The frequency difference of the modulator is proportional to the height, and can be processed to generate a current to drive a needle on a circular meter calibrated in feet, as in Figure 20.1, or digitised on a CRT, LED (light emitting diode) or LCD display, as shown in Figure 20.3.

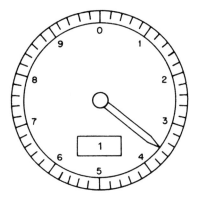

Figure 20.1   Radio altimeter.

The indication in Figure 20.1 is partly digital, in that the indication of thousands of feet is displayed in the central box of the instrument. This allows accurate depiction of height when close to the ground, which is where the radio altimeter is most useful. For the same reason, some indicators have a linear scale below 500 ft, with either a logarithmic or a different linear scale for heights above that, as in Figure 20.3. The method of processing the time delay actually produces 'steps' in the indication. These steps are at 5 ft intervals.

Some indicators can have their scale adjusted by the pilot. At high altitudes, each division may indicate 100 ft, at low altitude 10 ft. This is switched mechanically. Other radio altimeters do not indicate at all above about 2500 ft above the ground.

Accuracy

The radio altimeter has an accuracy of $\pm$ 3% of the indicated height, with a minimum error of $\pm$ 5 ft which is mainly due to the 'steps' in the indication. Because it is required to indicate most accurately when the aircraft is approaching to land, the indication is actually calibrated so that it reads the vertical distance below the wheels in the landing attitude. This means that when the aircraft is in the level flight attitude, or has its landing gear retracted, it will be higher above the ground than the instrument would suggest.

Decision height

A particularly useful feature of the radio altimeter is its ability to remind pilots that they are approaching their decision height on an instrument approach. Various warning systems can be fed by the radio altimeter, but the simplest is a marker on the face of the instrument itself. This can be combined with a light or audio warning, as in Figure 20.3. The marker is set by the pilot with the knob provided. In Figure 20.2, the decision height has been set and is shown on the display below the radio altitude, and a further warning, such as a light, an audio tone or often a recorded voice, is initiated when the two coincide. Most aircraft now incorporate the decision height warning into the GPWS or EGPWS, as described in Chapter 21.

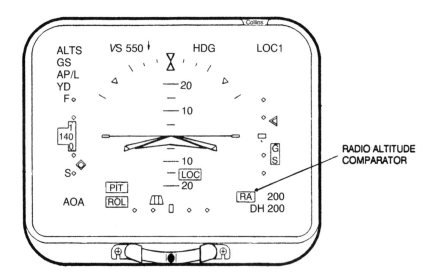

Figure 20.2 'Glass' display of radio altitude.

Radalt advantages

The radio altimeter has advantages over the barometric altimeter:

- It tells the pilot his height above the ground below him
- It does not require the input of pressure settings
- A wrong pressure setting does not affect it; in fact it does not suffer from any of the altimeter errors

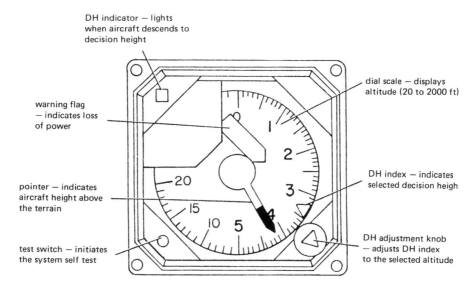

DH indicator — lights
when aircraft descends to
decision height

dial scale — displays
altitude (20 to 2000 ft)

warning flag
— indicates loss
of power

DH index — indicates
selected decision heigh

pointer — indicates
aircraft height above
the terrain

test switch — initiates
the system self test

DH adjustment knob
— adjusts DH index
to the selected altitude

Figure 20.3   Radio altimeter/decision height indicator.

Disadvantages

- It cannot provide vertical separation from other traffic under IFR using cruising levels
- It cannot provide height above a runway ahead
- It cannot provide any warning of the terrain ahead of the aircraft, only that which is directly below it

Altitude warnings

JAR-OPS 1.660 lays down that all new turbine-powered aeroplanes with a certificated maximum take-off mass above 5700 kg or more than 9 approved passenger seats must have an altitude alerting system. This system must be capable of alerting the flight crew upon approaching a selected altitude, and when deviating from that selected altitude. The alerts must be aural, and may also be visual. This is not the same as the decision height warning detailed above, because it relates to a pressure altitude, and usually takes digitised static pressure inputs from the air data computer.

In common systems on the market, the pilot selects his desired altitude with a knob on the display window of a selector on the glareshield or panel. A light on the selector, and another on the primary altimeter marked 'altitude alert', illuminate. The altitude alert unit monitors the display on the primary altimeter, or the output from the air data computer, and compares it with the altitude set on the selector. A selector is shown in Figure 20.4. The selector knob is at bottom right, the light at top right.

Figure 20.4   Altitude alert selector (courtesy Meggitt Avionics).

As the aircraft approaches the selected altitude, 900 ft before it reaches it an audio warning sounds. When 300 ft from the selected altitude, the lights go out. If the aircraft subsequently deviates by more than 300 ft from the selected altitude, the lights flash and the audio warning sounds. Once the aircraft deviates by more than 900 ft, the system resets itself. Such systems often form part of a reduced vertical separation minimum (RVSM) navigation system, required for flight in airspace designated as such.

Radalt summary

Principle	• Progressively changing CW reflected from ground below aircraft • Radalt measures time between Tx and Rx of each frequency = height
Frequency band	SHF (4200–4400 MHz)
Errors	Fixed error – steps of 5 ft
Range	Mainly to 2500 ft
Accuracy	• 5 ft ± 3% of indicated height • Set for landing attitude, not in taxiing attitude
DH	Decision height knob set by pilot gives light or audio indication of DH
Presentation	Dial, sometimes logarithmic above 500 ft, now on EFIS
Important	Radalt gives clearance below aircraft – not from terrain ahead

Sample questions

1. How does a radio altimeter work?

   a. It measures the Doppler shift of a CW transmission
   b. It measures the Doppler shift of a pulse transmission
   c. It measures the time from transmission to reception of CW frequencies
   d. It measures the time from transmission to reception of a pulse signal

2. What would be a typical wavelength of a radio altimeter transmission?

   a. 7 mm
   b. 7 cm
   c. 14 cm
   d. 1.4 m

3. The radio altimeter indicates 2000 ft. What is the minimum height that the aircraft could be?

   a. 1935 ft in the cruising attitude
   b. 1935 ft in the landing attitude
   c. 1898 ft in the cruising attitude
   d. 1899 ft in the landing attitude

4. Which of the following information will a radio altimeter give the pilot?

   a. His decision height
   b. His height above the ground ahead
   c. His actual height above the runway he is approaching
   d. All the above

5. How can radio altimeter information be presented to the pilot?

   a. On a dial with logarithmic scale
   b. By lights
   c. Aurally
   d. All the above

6. An aircraft is flying at FL 250 with its altitude alerting system set to that flight level. If it enters a climb, at what altimeter indication will the altitude alert warn the pilot?

   a. 25 100 ft
   b. 25 300 ft
   c. 25 500 ft
   d. 25 900 ft

Chapter 21
Terrain Avoidance Systems

Introduction

Controlled flight into terrain (CFIT) is the most common type of fatal accident to aircraft. Over recent years, many attempts to provide warnings to alert pilots to an imminent collision with the ground have been developed. The radio altimeter itself was an early attempt to reduce the problem. A Ground Proximity Warning System (GPWS), using inputs from the Radalt and other equipment on board the aircraft, became mandatory in the 1990s for large passenger aeroplanes. The improvement in accident rates as a result of fitting GPWS equipment led to its adoption for smaller aeroplanes also.

False alarms and the inability of the inputs to identify terrain ahead of the aircraft were two limitations of a GPWS which required improvement. The search to eliminate or reduce these problems led to the development of EGPWS (Enhanced GPWS) using a computer database containing terrain structure, and a GPS receiver to provide warning of trouble ahead. The main capabilities of this enhanced system were encapsulated into an ICAO requirement for a terrain awareness and warning system (TAWS), which is becoming a mandatory requirement for all medium and large transport aeroplanes at the time of writing.

GPWS

Principle

The GPWS takes inputs from aircraft sensors and navigation equipment and feeds them into a computer, or central processing unit (CPU). The CPU compares the inputs, decides which are the most important for the phase of flight, and provides two levels of warning to the pilot. The first level, that of an 'alert', tells the pilot that he is approaching a dangerous situation, and should take action to return to a safe condition. The second, more critical stage, is a 'warning', which tells the pilot he is already in a dangerous situation, and must take an immediate, standard, action to remedy it.

GPWS inputs

The radio altimeter gives a constant feed of height above ground to the CPU. A vertical speed sensor, digitising the VSI information (in fact taken from the air data computer (ADC)), is also constantly providing an input. The position of the landing gear and that of the flaps or other high lift devices, is also provided. The ILS glidepath is the last input which the CPU uses.

GPWS outputs

If the CPU decides that there is a potential danger of colliding with terrain, it will put out warning signals to the pilot in both visual and audible forms. The CPU will also put out indications of computer failure and any failures which may occur in one of the input signals, to a monitor indicator. A block schematic diagram of the inputs and outputs is shown in Figure 21.1 below.

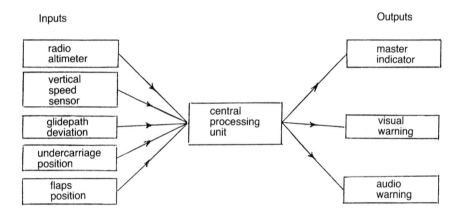

Figure 21.1   Elements of GPWS.

GPWS modes

Mode 1   An excessive rate of descent when the radalt is indicating less than 2500 ft provides the first alert. The alert uses the words 'Sink rate, sink rate'. The warning uses the words 'PULL UP, PULL UP', usually coupled with a 'whooping' noise and a flashing lamp.

Mode 2   Depends on the same inputs as Mode 1, with the addition of the position of the undercarriage and flaps. If the rate of descent is high below a radalt height of 1800 ft with gear and flap up, the alert Mode **2A** is given. The same alert, but described as Mode **2B**, is given if the high sink rate occurs below 790 ft, even with gear and flap in the landing configuration. The alert uses the words 'Terrain, terrain', and the warning is the same as in Mode 1.

Mode 3   Applies after take-off or go-around, when the aircraft has started climbing and gear and/or flap has been selected up. If the aircraft subsequently descends (sinks), the alert sounds. The alert words are either 'Don't sink, don't sink' or 'Terrain, terrain'. The warning is again the same as Mode 1.

Mode 4   An alert sounds if the aircraft is below a certain height without being in the landing configuration. Mode **4A** applies below 500 ft with the gear up, Mode **4B** if below 200 ft with the flap up. Mode 4 alerts use the words 'Too low – Gear/Flaps' (as appropriate). The warning for Mode **4A** is the same as Mode 1, but for Mode **4B** the warning is 'Too low – terrain'.

Mode 5   Uses ILS glidepath information. If the aircraft descends below the glideslope to such an extent that the instrument gives more than $\frac{1}{2}$ fly up demand, Mode 5 alerts the pilot with the word 'Glideslope'.

Mode 6   Is straightforward information that the aircraft has descended below the selected decision height, and uses only the radalt input. The word used for the alert is 'Minimums'.

Some older equipment may only provide warnings for Modes 1, 2, 3 and 4, and alerts for Mode 5.

Pilot's actions

On receiving a GPWS 'Alert', which will be the words as detailed above, the pilot must take action to get out of the dangerous situation and stop the alert indication. That action may be to reduce his rate of descent, level off, turn to a safer area, or even to lower his gear or flap.

If he receives a 'Warning' however, with the 'whooping' sound and the words 'PULL UP', he has no options, unless he is in good weather and absolutely sure the warning is a false alarm, in which case he is permitted to treat a warning as an alert. Otherwise, he must treat the 'warning' as a command to immediately level the aircraft's wings, apply take-off power, and climb away at the speed and angle for maximum gradient climb until the aircraft has reached the en-route minimum safe altitude.

Requirements

JAR-OPS 1.665, converted into UK law by the Air Navigation Order, requires that all turbine-powered aeroplanes with a maximum certificated take-off mass over 5700 kg or a maximum approved passenger seating configuration of more than 9 must be equipped with a ground proximity warning system. That system must automatically provide 'timely and distinctive warning' to the crew of sink rate, ground proximity, altitude loss after take-off or go-around, incorrect landing configuration and downward glideslope deviation. This warning must be given by aural signals, which may be supplemented by visual signals.

Serviceability

The equipment has a built-in self-testing system, (BITE or built-in test equipment), which monitors the GPWS serviceability. When selected, which can only be done on the ground, it displays the visual and audible warnings simultaneously.

If the GPWS is unserviceable, the aircraft should not fly. However, if it is not possible to repair the equipment at the aerodrome where the unserviceability is discovered, it is permitted for the aircraft to continue its flight to an aerodrome where it can be repaired, provided that flight consists of no more than 6 segments.

The system is also monitored during flight. Any failure is automatically indicated on the flight deck.

GPWS problems

There were some false alarms generated by the GPWS equipment, usually while the aircraft was following approach procedures in hilly terrain. However, the major problem with GPWS is that the system has no knowledge of the terrain ahead of the aircraft; separation can only be measured from the ground directly beneath the aircraft. It was necessary to add a predictive terrain hazard warning function to the system.

TAWS

The advent of navigation systems which could provide great accuracy, coupled with powerful computers which process large amounts of information, allowed the development of systems which could provide such a predictive terrain hazard warning function (warning of terrain ahead of the aircraft). The first TAWS on the market, called the enhanced ground proximity warning system (EGPWS), can also give warning of excessive bank angle and windshear on the approach.

TAWS compares the aircraft position in three dimensions with that of the ground around it. It is programmed to provide warnings in a similar fashion to GPWS, and displays the information for the pilot if a suitable screen is available It also provides standard alerts and warnings as a normal GPWS.

TAWS inputs

The radio altimeter provides information to determine the aircraft's vertical position above the ground. Continuous 'raw' information from the navigation computer position is used to determine the aircraft's horizontal position and track. As with GPWS, the aircraft's air data computer (ADC) provides details of the aircraft's vertical and horizontal speed and rates of change, but also provides static air temperature. ILS information comes from the VHF navigation receiver, and the flap and gear positions are also fed in. The FMS, inertial reference system (IRS), or attitude, heading and reference system (AHRS), provide attitude, track angles, headings, and groundspeed. If an IRS is not available, accelerometers can provide accelerations in all dimensions. An angle of attack (AoA) or stall warning system also provides information.

The CPU compares all that information with the radio altitude as in GPWS, and also with its databases of the ground profile in the area towards which the aircraft is flying. These databases are provided for general terrain worldwide, for obstacles in certain areas, and for all major aerodromes in progressively more accurate form.

TAWS legislation

The importance of reducing the number of CFIT accidents has led to rapid changes in legislation. ICAO now requires all transport aeroplanes to install and use serviceable TAWS equipment. The UK has accepted that aeroplanes which already have GPWS equipment installed may delay fitment of TAWS equipment until January 2005. However, all turbine aeroplanes over 5700 kg that are authorised to carry more than 9 passengers, and which do not have GPWS already installed, must carry a serviceable TAWS from 1 October 2001 (public transport aircraft must have it from 1 January 2001).

In a similar fashion to GPWS, the aeroplane must not fly unless the TAWS is serviceable. However, if it is not possible to repair an unserviceable TAWS at the place where it becomes unserviceable, the aircraft may fly to a suitable repair base. Again, this concession only applies for up to 6 route legs.

TAWS alerts and warnings

The basic alerts and warnings are the same as GPWS, although the modes may be defined slightly differently, and expanded as shown below. Amber lights are usually used to give visual indication of alerts (also sometimes called 'cautions') and red lights give warnings.

Mode 1 Is the same as GPWS, in that it gives alerts and warnings of excessive sink rate below 2500 ft above ground.

Mode 2A and 2B Again use radio altimeter and configuration information to alert and warn of reducing height above the terrain below, as in GPWS.

Mode 3 Again provides alerts and warnings of decreasing height above terrain after take-off or missed approach.

Mode 4 Provides alerts and warnings of insufficient terrain clearance if the aircraft is not in the landing configuration. Mode **4A** alerts of descent below 500 ft with gear up. Mode **4B** alerts of descent below about 250 ft with flap up. Mode **4C** adds alerts if the radio altimeter reading decreases below a certain proportion of its previous value after take-off. The actual height values at which the particular alerts are triggered vary with airspeed.

Mode 5 Is again related to the ILS glideslope. The commercially available EGPWS provides two levels of alerting. A so-called 'soft' alert, 'Glideslope' at half volume with the 'caution' light, is given when below 1000 ft radio altitude and more than 1.3 dots below the glideslope. As the deviation below the glideslope increases, the alert becomes progressively more frequent. When $2\frac{1}{2}$ dots below the glideslope (the $\frac{1}{2}$ fly-up command), the alert becomes 'hard', and 'GLIDESLOPE' is heard at full volume, signifying the change to a 'warning'. Below 150 ft, the alerts are desensitised. In EGPWS, Mode 5 alert bands vary with rate of descent and can be inhibited automatically for backbeam approaches or manually by the crew.

Mode 6 Is a programmed warning facility giving voice 'callouts' at selected radio altimeter readings. Apart from the call of 'Minimums' or similar, heights in numbers of feet are called out. Other calls can be 'approaching minimums', or similar. Various tones can be played over the audio system as different heights are passed. In EGPWS, Mode 6 also provides alerts of excessive bank angle. The limit set varies as radio altitude, reducing rapidly below 150 ft. The alert consists of the words 'Bank angle, bank angle'.

Mode 7 If fitted in EGPWS, is designed to provide alerts if the aircraft encounters windshear below 1500 ft. A typical windshear warning addition to a TAWS is described later.

TAWS additions

The 'look ahead' alerting facility of a TAWS projects the aircraft flight path forward and compares its future position with the available databases. When terrain or obstacles infringe the calculated flightpath at calculated possibly hazardous ranges (the actual range depends on speed), the caution lights illuminate and the words 'caution, terrain/obstacle' (as appropriate), or similar, are spoken. As the aircraft closes to the confliction, the alert changes to the warning light and the words become 'TERRAIN, TERRAIN (or obstacle, obstacle), PULL UP, PULL UP!'

The various databases available provide different levels of protection in different areas. The main database of terrain structure is a model of the surface of the earth. Of necessity, it is a rough model, and contains only the general contours. Safety is provided by 'rounding up' the smaller variations in contours.

More detailed databases can be provided around planned routes. These can take into account known obstructions, as well as a more detailed contour model. Around known aerodromes, the system assumes aircraft may wish to descend, and gives less warning of terrain proximity. However, away from these known aerodromes at increasing ranges, the safe 'buffer' provided becomes greater.

Finally, databases are also provided of the terrain and obstacles under the approach and missed approach paths of these aerodromes.

EGPWS display

The CPU decides what terrain is likely to encroach on the aircraft's flight path, and displays it on the EFIS screen, or on an AWR screen if available and EFIS is not fitted. This may be termed the TAD (terrain awareness and alerting display). Terrain is shown in shades of red, yellow and green (for 'danger', 'caution', and 'safe but present', rather like the AWR display).

The normal display on a typical system shows all terrain above the aircraft and less than 2000 ft below it. The terrain is shown out to the range selected on the display. At long range, terrain more than 2000 ft above the aircraft is shown as red dots. Between 2000 and 1000 ft above the aircraft is shown as close yellow dots, between 1000 ft above and a programmed distance below the aircraft is shown in widely separated yellow dots. Green dots show terrain from there to 1000 ft below the aircraft, and widely separated green dots shows ground between 1000 and 2000 ft below.

When the EGPWS decides that an alert should be given (about 1 min from collision), the terrain which caused the alert changes to solid yellow. An EGPWS warning (30 s from impact) generates a solid red image, as seen in Figure 21.2.

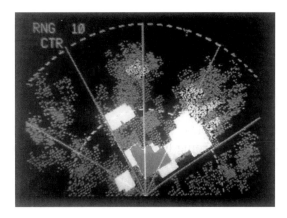

Figure 21.2   Terrain alerting display (courtesy Honeywell).

Pilot actions

Generally, if the pilot receives a TAWS alert he should stop any descent and climb as necessary to eliminate the alert. Then he should analyse all available instruments to determine the best course of action. At a suitable point, he should advise ATC of the situation.

If he receives a TAWS warning, he must apply maximum available power, disengage the autopilot, and 'smoothly but aggressively' increase pitch attitude towards the 'stick shaker' or pitch limit indicators, to obtain maximum climb angle. The climb must be continued until the warning is eliminated and safe flight is assured, advising ATC of the situation.

Specific actions in the case of soft 'glideslope' alerts are to return to the correct glideslope. A hard 'glideslope' alert indicates that the aircraft has descended below the safe angle (see Chapter 8, ILS). The pilot must treat it as a warning, and a missed approach must be flown.

EGPWS self-test

The BITE (built-in self-test) system has various levels to check different parts of the system. Pilots will normally initiate a basic self-test as a preflight check. When the test switch is made, the system should check all the warnings and alerts in sequence by sending simulated signals through each of the inputs, and producing the correct instrument, aural and visual signals.

Windshear alerts and warnings (EGPWS)

The EGPWS as available from Honeywell has a windshear detection system as described in Chapter 12. It compares airspeed information from the air data computer (ADC) with groundspeed information from the navigation computer to detect changes in wind affecting the aircraft, and likely wind-shear turbulence.

The warning computer will provide windshear cautions (alerts) if an increasing headwind (or decreasing tailwind) and/or a severe updraught exceed a defined threshold. These are characteristics of conditions which might be expected just before an encounter with a microburst (see *Ground Studies for Pilots: Meteorology*). The caution light appears and the words 'Caution, Windshear', or 'Windshear ahead' are spoken.

Windshear warnings are given if a decreasing headwind (or increasing tailwind) and/or a severe downdraught exceed a defined threshold. These are characteristic of conditions which might be experienced within a microburst itself, or just afterwards. The warning consists of the words 'WINDSHEAR, WINDSHEAR, WINDSHEAR' and the illumination of the red warning lights.

The alert and warning thresholds depend on available climb performance, flight path angle, airspeed changes, and fluctuations in static air temperature. Windshear warnings take priority over all others, but windshear alerts are low priority.

Other TAWS equipment

The cost of EGPWS militates against its use in general aviation aircraft, so manufacturers are working on systems which cost less but still allow some measure of warning of possible terrain confliction. Other systems are also under development by various manufacturers to provide TAWS using totally different concepts. One of those involves developing current military equipment to provide infra-red imaging of terrain ahead on a head-up display (HUD).

GPWS summary

Principle	• Takes inputs from aids to warn pilot of hazards near the ground • Central processing unit (CPU) gives visual and audible signals
Inputs	• Radalt • Vertical speed sensor • ILS glidepath • Undercarriage position • Flaps position

Modes	1.	Excessive rate of descent below 2500 ft
	2A.	High sink rate below 1800 ft, no gear or flap
	2B.	High sink rate below 790 ft, landing configuration
	3.	Sink after take-off or go-around
	4A.	Below 500 ft – gear up
	4B.	Below 200 ft – flap up
	5.	$\frac{1}{2}$ fly-up indicator on ILS
	6.	Descent below minimums
Alerts	Warning of possible hazard, pilot must take action to remedy situation	
Warnings	Pilot must level wings and initiate max gradient climb to safe altitude. (Unless good visual meteorological conditions (VMC) and obviously not in danger)	
Test	BITE preflight	
Unserviceability	Aircraft may fly until it lands at an airport where repair is possible. Maximum of 6 legs	
Important	GPWS gives clearance below aircraft – not from terrain ahead	
TAWS (Enhanced GPWS)	• GPS position compared with database • Terrain contours ahead presented on nav displays – Red/Yellow/Green • Alert given 1 min before impact (GPWS: 10–30 s)	

Sample questions

1. What does a GPWS Mode 3 alert indicate to a pilot?

   a. The aircraft is close to the ground with the undercarriage up
   b. The aircraft is sinking after take-off or go-around
   c. The aircraft is below its selected decision height
   d. The aircraft is approaching high ground

2. Which of the following is not an input to a GPWS?

   a. Radio altimeter
   b. ILS localiser
   c. Vertical speed sensor
   d. Flap position sensor

3. Which of the following would not trigger an alert from the GPWS?

    a. A high sink rate at 1000 ft agl in the landing configuration
    b. A high sink rate at 1500 ft agl with gear and flap up
    c. A low sink rate during go-around
    d. Level flight at 400 ft with undercarriage up

4. How low on the ILS glideslope will the GPWS alert the pilot?

    a. When the aircraft is below the bottom of the glideslope
    b. When the aircraft is more than 3° below the glideslope
    c. When the ILS indicator shows more than half a full fly-down indication
    d. When the ILS indicator shows more than half a full fly-up indication

5. Which mode of the GPWS will warn the pilot when he descends below 500 ft with his gear still up?

    a. Mode 1
    b. Mode 2b
    c. Mode 3
    d. Mode 4a

6. Below what height will GPWS Mode 1 warn the pilot of a high rate of descent?

    a. 2500 ft
    b. 1800 ft
    c. 790 ft
    d. 500 ft

7. Regulations require a pilot to level his wings and initiate a maximum gradient climb to safe altitude under certain conditions. What are these conditions?

    a. Any time he receives a GPWS warning
    b. Any time he receives a GPWS alert
    c. If he receives a GPWS alert, unless in VMC and obviously in no danger
    d. If he receives a GPWS warning, unless in VMC and obviously in no danger

8. If GPWS is found unserviceable, it must be repaired as soon as possible. If the FBO at destination has not got the necessary facilities, what must the pilot do?

a. He must divert to an airfield with repair facilities

b. He may land at destination, but must wait for the GPWS to be repaired

c. He may land at destination, but must make the next landing at a repair base

d. He may continue with his planned schedule until he lands at a repair base

9. How does a pilot know his GPWS is unserviceable?

a. ATC will tell him

b. A warning light will illuminate when he descends below 2500 ft

c. He will test it before flight

d. He has no means of knowing

10. Which of the following circumstances will GPWS not give warning about?

a. A high rate of descent close to the ground

b. Rising ground ahead

c. Approaching with the undercarriage not locked down

d. Approaching without flaps lowered

11. What is the principle of EGPWS?

a. It combines the radio altimeter with many sensors to compute safe descent rates at varying heights

b. It sends radar signals ahead of the aircraft to sense high ground ahead

c. It gives an infra-red TV picture of the flight path through cloud

d. It uses GPS and a map database to compute safe clearance from ground ahead

Chapter 22
Hyperbolic Principles and LORAN-C

Introduction

For a long period during and after the Second World War, and before the advent of satellite navigation, navigation aids using the hyperbolic principle were the mainstay of long distance navigation. Using MF, LF and VLF frequencies, Decca, Loran, and Omega with their associated computers were able to produce fixes over any part of the globe.

Decca chains were withdrawn as the millennium ended. Omega ceased transmission in 1997. Only LORAN-C (LOng RAnge aid to Navigation) remains at the time of writing, reprieved primarily as a back up for satellite navigation systems by the United States and Russia. However, the other systems remain in certain examination syllabi.

Hyperbolic principle

If signals are sent from two places at the same time, and a receiver can receive both of them, the difference in time between receiving the two signals will indicate a position line. For example, if the signals are received together, the receiver must be an equal distance from each station. Its position is somewhere on a line which lies half-way along and at 90° to the line joining the two stations (the baseline). If one signal were received before the other, then the receiver must be that distance (time multiplied by speed of radio waves) closer to the station whose signal was received first than to the second station.

The shape of this position line is a hyperbola. In Figure 22.1 the hyperbolic line QPR equates to a distance difference of 4 nm. Every point on the line QPR is 4 nm closer to station S than to station M. The line MS is the baseline, and the line BAC, the 'bisector line', is the hyperbola of zero difference. Each time (or distance) difference line is unique, and the pattern of these lines forms an unambiguous hyperbola family.

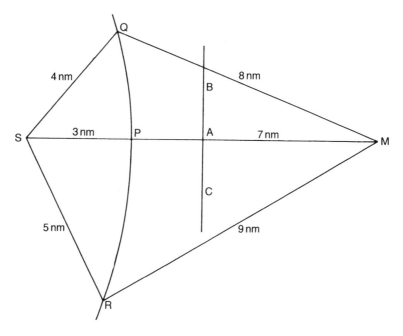

Figure 22.1   Hyperbolic position line.

LORAN-C

Principle

To simplify receiver technology, the two stations do not transmit together, but one after the other. The 'master' station transmits first and the 'secondary' or 'slave' station transmits when it receives the master's signal. There is therefore a delay between the two transmissions, the 'master/secondary propagation delay'. This delay is not long enough to allow use at anything other than short ranges, so a further delay is built in by each secondary transmitter, the 'secondary specific delay'. The airborne equipment subtracts those known fixed intervals from the total 'propagation delay' to determine the time difference and calculate the distance difference between the two stations. By using one master and several secondaries transmitting with varying specific delays, more position lines can be produced to give a fix.

Transmissions

Like earlier systems, LORAN-C uses 'chains' of stations to provide fixing coverage. Each chain consists of one master and a group of up to four secondaries, transmitting with different 'secondary specific' delays. The secondaries are designated W, X, Y, and Z in the order of their transmissions. Some chains may not use the W slave, and a chain with only two secondaries will dispense with the Z slave also.

Every chain transmits on the same frequency, 100 kHz. This low frequency means that the receiving aircraft does not need to be in line-of-sight of the transmitters, in fact the receivers are designed to work mainly with ground waves. It also allows the signals to be received at quite long ranges. The chain baselines are typically about 600–1000 nm long.

The signals are sent in groups of pulses. This allows the airborne equipment to add them together and effectively increase the power of the signal received in relation to the background noise. To distinguish between chains each chain has its own 'group repetition interval' or GRI, which is the time between the first pulse of each group. This GRI is used to designate the particular chain with letters and numbers arranged in a code to show the exact GRI of the chain. The letters show a basic GRI, and the numbers show the number of hundreds of microseconds less than that basic difference which the chain actually uses.

$$S \quad = 50\,000\,\mu s$$
$$SH = 60\,000\,\mu s$$
$$SL = 70\,000\,\mu s$$
$$SS = 80\,000\,\mu s$$

Hence a chain designated SH7 would have a GRI of 53 000 μs, one designated SL0 a GRI of 70 000 μs, and one designated S3 a GRI of 47 000 μs.

The master station transmits a group of 8 pulses 1000 μs apart, followed by a further pulse 2000 μs later. All the secondaries in the chain receive the signal from the master, and after their individual secondary-specific delay, the slaves re-transmit the 8 pulses. The extra pulse allows the receiver to identify which signal comes from the master, and can also be modulated to carry specific information, such as the non-availability of a specific secondary.

Obtaining position lines

The airborne equipment matches the secondary transmissions with those from the master. Once matched, the time delay is measured. The secondary-specific delay is subtracted from the total propagation delay to produce the actual time difference and therefore the relevant position line. To minimise interference to users of adjacent frequencies, the transmitted pulses build up and decay in strength slowly. Matching is therefore not easy, especially as it originally had to be done manually.

To make matching easier, and to enable modern equipment to work automatically, the phases in the various pulses of each group are switched in a fixed pattern for the master and each secondary. The equipment can therefore determine which signal it is receiving at any time, and disregard any signals which are incorrect. These incorrect signals may be signals which are missing a pulse, or be long-hop sky wave signals from another Loran chain. The phase switching permits accuracies of $\pm 10\,\mu s$.

Sky waves at the Loran frequency of 100 kHz are strong. At very long ranges, up to about 2000 nm, the system can actually use them to provide less accurate fixes beyond the intended specification. However, at shorter ranges, they interfere with the ground waves. The system gets round this problem by using the third cycle of each pulse for matching, before the sky wave can reach the receiver. This is called 'indexing'. Originally, all this had to be done manually on a cathode-ray tube, but modern equipment is all computerised, and the calculations are automatic.

Accurate position lines using ground waves can be achieved at ranges up to 600 nm over land, and about 1000 nm over the sea. At ranges between that and about 1400 nm, the airborne equipment may only be able to receive one of the signals, either master or secondary, as a ground wave. If the missing signal can be received as a sky wave, it is still possible to produce a position line, although accuracy will be reduced. Because the sky wave has taken longer to travel to the receiver, a correction must be made to its timing. An extra delay, of $40\,\mu s$ by day or $60\,\mu s$ by night must be subtracted from the timing of whichever signal has been received as a sky wave.

Certain values of time difference are always found at certain positions in each chain. These are shown in Figure 22.2. For example, if the total difference is exactly equal to the 'secondary-specific' delay, then the receiver lies along the 'extended baseline' beyond the secondary transmitter S. If the total difference is equal to the sum of the 'secondary-specific delay' plus the 'master/secondary delay', then the receiver lies along the bisector AB. If the total delay equates to the sum of the secondary-specific delay plus twice the master/secondary delay, the receiver lies along the extended baseline beyond the master transmitter M.

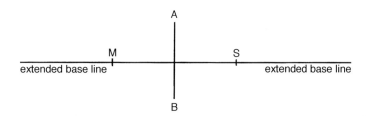

Figure 22.2 Specific positions.

Fixing

Sample hyperbolic position lines are plotted on special charts for each chain, and the operator can interpolate between those already plotted to produce his own position line. Combining two or more position lines from the master and different slaves produces a fix, which then has to be transferred on to a normal plotting chart. However, modern computers can convert the time differences electronically into latitude and longitude, doing away with the need for the special charts and saving time and effort.

Loran-C chains are positioned in the Pacific area, the Mediterranean, Arabia, Northern Europe and the North Atlantic. Some of these chains are expected to remain in service until at least 2008.

Displays

As stated earlier, original airborne equipment used manual slewing (adjusting the timebase) on a cathode-ray tube. Later automatic equipment displayed a digital time difference readout. However, modern equipment can provide a direct input to a navigation computer and display position in any way desired by the crew.

Accuracy

Traditionally, the LORAN-C accuracy has been accepted as ± 1 nm at a range of 1000 nm from the baseline. It is advertised at ± 0.25 nm by the current system operators.

European chains

The US Coastguard developed the original LORAN-C chains, but withdrew support from those outside the United States in 1994. Several European nations formed NELS (North-west European LORAN-C System) as a consortium to establish a new chain structure and to run the system in Northern Europe.

Chayka

A similar system called Chayka was developed by the Russians during the Cold War. It is compatible with LORAN-C, and the Russian chains provide an extension into eastern Europe. They are expected to remain in service until at least 2010.

Uses of the LORAN-C signal

The members of the NELS consortium propose to use LORAN-C not only as a stand alone aid, but in addition as one input to a duplex navigation system which also takes satellite navigation signals as the other input. It is also proposed to use the LORAN-C transmissions to carry DGPS corrections on their ground wave signals.

Summary

Principle	• Differential range by pulse timing • Master sends coded pulse group, slave delays, retransmits • Time difference gives hyperbolic line • Indexing provides freedom from sky wave interference
Frequency band	LF (100 KHz)
Range	2000 nm
Accuracy	1 nm or better at 1000 nm. Less accurate sky wave positioning
Failure	Chain transmits warning signals
Display	• Old – CRT or time difference readout • New – computerised readout of Lat and Long etc.
Coverage	Pacific, Mediterranean, Arabia, N Atlantic
Chayka	Russian equivalent to LORAN-C

Sample questions

1. What does a LORAN-C receiver measure?

   a. The phase difference between 2 signals
   b. The difference in frequency between 2 signals
   c. The difference in strength between 2 signals
   d. The difference in time between 2 signals

2. How do the transmitters of a LORAN-C system operate?

   a. Stations send pulses at different start times
   b. Stations send pulses at the same time

   c. Stations send continuous waves, starting at different times

   d. Stations send continuous waves, starting at the same time

3. What would be a suitable wavelength for a LORAN-C signal?

   a. 3 km

   b. 300 m

   c. 30 m

   d. 3 m

4. Which of the following is true about the approximate range and frequency band of LORAN-C?

   a. Range is 2000 km, band is MF

   b. Range is 2000 km, band is LF

   c. Range is 4000 km, band is MF

   d. Range is 4000 km, band is LF

5. What is the quoted accuracy of LORAN-C?

   a. 1% of range up to 1000 nm, less accurate from sky waves

   b. 1 nm up to 1000 nm, less accurate from sky waves

   c. 1% of range up to 1000 nm, no sky waves are used

   d. 1 nm up to 1000 nm, no sky waves are used

6. If a LORAN-C signal fails, which of the following is true?

   a. There is no warning

   b. The instrument shows a warning flag if no signal is received

   c. The stations transmit warning signals to prevent false indications

   d. The instrument switches to memory mode

7. LORAN-C consists of chains of master and secondary stations. Which of the following is correct?

   a. Each secondary retransmits the master's signal immediately it receives it

   b. Each secondary sends an independent signal at a fixed time

   c. Each secondary retransmits the master's signal after the same delay

   d. Each secondary retransmits the master's signal after individual delays

8. If the total propagation delay between master and secondary is equal to the sum of the secondary-specific delay and the master/secondary propagation delay, what can be said about the position of the receiver?

   a. It is on the extended baseline in the direction of the master

b. It is on the extended baseline in the direction of the secondary
c. It is on the bisector line
d. It is above the master station

9. Why is cycle matching carried out at the end of the third cycle of LORAN-C pulses?

a. To identify individual chains
b. To reduce sky wave interference
c. To provide the maximum signal/noise ratio
d. To avoid confusion between master and secondary signals

Chapter 23
Obsolete Hyperbolic Systems – Decca and Omega

Introduction

The Decca navigator system used the principle of hyperbolic position lines (as explained in the last chapter), but in a different fashion. Ground stations sent continuous wave signals in the LF band which were phase compared to provide accurate navigation for ships. It was also sold as an airborne navigation aid, and was accepted as an alternative to ADF and VOR for flight in UK controlled airspace. It had the one advantage that it was not a line-of-sight aid, but because it used surface waves it could be used down to sea level.

With the advent of satellite navigation systems (see Chapter 15), a simpler system was available with the same advantage. Most of the Decca chains around the world were closed down by 2000. However, the equipment remains, and the system may be resurrected. At the time of writing it remains in the syllabus for JAA professional pilot examinations.

Principle

Phase comparison

Two continuous wave transmitters on the same frequency are spaced at a whole number of wavelengths apart. A receiver will receive the signal from both, but at different phases. Transmissions are phase locked together, so that each one starts with the same phase at the same time. Subtracting the phase of one from the phase of the other will give a phase difference, as against the time difference used in LORAN. The lines of constant phase difference are hyperbolic position lines. The phase difference can be shown on phase meters in the aircraft, or fed to a computer processing unit.

In practice, if the transmissions were on the same frequency, the receiver would be unable to differentiate between them. The transmissions are in fact on different frequencies, and a processor multiplies each of them by a sufficient amount to bring them up to a common frequency. The phases can

Figure 23.1 Out of phase signals.

then be compared. Figure 23.1 shows two signals at that stage, prior to comparison.

The chain

Each Decca chain consists of 4 transmitters, a master and 3 slaves. The slaves should ideally be placed in a triangular pattern with the master at the centre, although the angles are not critical. The slaves have names, red, green and purple, corresponding to the colours of the corresponding position lines on a published Decca chart. Each chain has its own base frequency, and each station transmits on a multiple of that base frequency f. The master always transmits on $6f$, the Red slave on $8f$, the green slave on $9f$ and the purple slave on $5f$. For example, if the base frequency was 14.1 kHz, the master would transmit on 84.6 kHz, the red on 112.8 kHz, the green on 126.9 kHz, and the purple on 70.5 kHz. All the frequencies can be said to be 'harmonics' of the base frequency, as the base frequency is a 'harmonic' of each transmitted frequency.

The airborne equipment receives all the transmitted frequencies, and compares them at frequencies equating to the lowest common multiple of each combination. For example, the comparison between the master and the red would be at $24f$, which in the above example with a base frequency of 14.1 kHz equates to 338.4 kHz. LF signals have the ability to travel long distances, so each chain uses a difference base frequency to avoid interference from others.

Lanes

The phase comparison method gives accuracy, but unfortunately considerable duplication of possible position lines. For instance, along the baseline, the phase of each signal changes through 360° every wavelength, but the phase difference varies through 360° every half-wavelength. Each time this cycle occurs, another possible position line can be plotted. We can only be sure of the position within a 'lane' corresponding to one cycle (equating to one half-wavelength of the comparison frequency at the baseline). Lanes are therefore different widths. Each lane starts at the position line giving zero phase difference, and continues to the next zero difference. Each lane is given a number, starting from the lane nearest the master station. Rather than use degrees, the phase changes are usually referred to as decimals of the lane number.

On a ship, having lanes about 400 to 600 yd wide is no major problem, because it moves slowly and navigators can plot positions regularly as the ship moves across lanes. In aircraft, it is vital to be able to differentiate between lanes. A computer with continual information from the stations can avoid 'lane slip' (missing a complete lane as the aircraft flies across lanes). However, older equipment relies on the navigator, and at high speed keeping track of the lanes is very difficult.

Lane identification

To allow for lane identification, the signals are interrupted and extra pulse transmissions on different harmonics are made by each station. Together with their normal transmissions, the master and each slave in turn transmit two frequencies which can be subtracted to give that base frequency. The basic lane identification system compares the phases of the base frequency to give a coarse position line which indicates the lane in which the accurate phase difference lies.

This system does not totally identify the lane. Even at the base frequency, the phase changes occur at half that wavelength, which corresponds to a considerable distance but only a small portion of the whole pattern of hyperbolic lines. We are left with 'wider zones', which each contain a number of lanes depending on the slave; 18 for the green, 24 for the red, and 30 for the purple. Zones are of the same width. Each zone has a letter, starting at the extended baseline through the master, so a position line obtained from a chain master and slave is referred to by the colour, a letter, a number and decimals. The lane numbers repeat in each zone, the red starting with 0, the green with 30 and the purple with 50.

Zone identification

Without some means of identifying the zones, the navigator must set up the receiver equipment at least in the correct zone in which it is situated. Navigation computers fed by the Decca chains must have the dead reckoning position loaded before use.

The later Decca Mk 10 system uses a more complex system of pulses transmitted on all harmonics at various times. This is called a 'multipulse' system. It includes an extra frequency $(8.2f)$ which is compared with the $8f$

signal transmitted at the same time to give a fraction of the base frequency (1/5). This has an effective wavelength equal to the baseline, so the resultant phase comparison pattern becomes two large areas, each including five original zones. Confusion between these areas can still exist, so a deduced reckoning position must still be fed into the airborne equipment when it is switched on.

Airborne equipment

Decometers

The original system of presenting the information was on three phase meters, each referred to by the colour of the relevant slave. These 'decometers' had a digital counter for the zone letter and an analogue face for the lane number and the decimals. The information had to be transferred to a chart marked with the hyperbolic lattice. On switching on the lanes had to be manually set. A decometer is represented in Figure 23.2. Lane identification was provided by a fourth decometer with a 'spider' pointer which indicated six 'vernier' positions at a time. When identification was taking place, a light would shine in the relevant decometer, and a coarse 'sector' marker on the fourth meter showed which of the six verniers referred to the lane being identified at the time. A lane identification meter is shown in Figure 23.3.

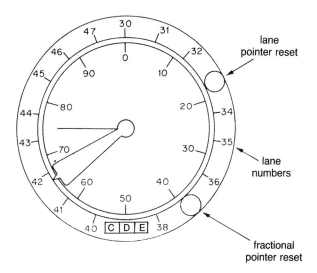

Figure 23.2   Decometer indicating D 41.75.

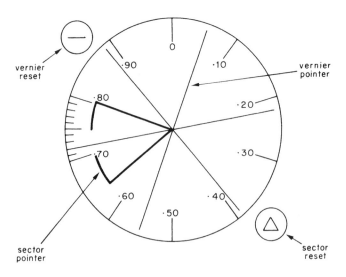

Figure 23.3   Decometer lane identification meter.

Decca flight log

To provide some simpler method of displaying information without having to transfer position lines, the Decca company produced a series of simple drum maps covering routes requested by customers. A pointer driven by the decometer signals moved over this 'flight log'. The map display was distorted, but distortion was at a minimum close to the intended route. Later flight log systems were able to dispense with the decometer display.

Locked oscillators

During identification, the navigation signal for the relevant master/slave pair was discontinued, and this encouraged lane slip (see below). More modern receivers introduced oscillators into the receiver, which continue to produce a signal at the correct phase on the signal frequencies even when lane identification is in progress. The oscillator 'locks on' to the phase of the original received signal, and thereafter produces the signals actually used for processing. The external signals are only used to update the oscillators.

This technique reduced lane slip considerably, and made possible the multipulse system in Decca Mk 10 and later equipment. Old chains, called 'V' chains, were superseded by the multipulse chains, but not all were replaced.

Digital display units

Computers have the capacity to process the Decca signals directly into latitude and longitude, which can be displayed digitally on drums or screens. The original digital display unit was used in addition to the decometers. The display could be 'frozen' while a fix was taken and afterwards would reset itself to the correct position.

Decca R-Nav

The latest systems will select the nearest Decca chain and tune it in automatically. Having tuned it, they will lock on and decipher the Decca signals to give latitude and longitude position, and compute track and ground speed. They will stay with the chain in use until another chain provides a notably better geometry. They can use either multipulse or 'V' chain signals. The calculated position can be compared with intended future waypoints, and the systems can display directions for the aircraft to follow on a course direction indicator (CDI) or on an EFIS display.

Ground equipment

Chain identification

Chains are often recognised by their names (e.g. the English chain). However, each chain is identified by a number and letter code, corresponding to one of the 73 base frequencies. The numbers go from 0 to 10, and the letters from A to F. Frequency 0A corresponds to a base frequency of 14.0167 kHz, which has a master frequency of 84.1 kHz. Frequency 10C, the highest, corresponds to a base frequency of 14.3175 kHz, a master of 85.905 kHz.

Stations

Each station is triplicated; three transmitters are positioned at each. One is transmitting, a second is on standby in case the primary transmitter fails, and the third is available for maintenance. Power supplies are also available in case of problems. Standby generators are kept available, as are batteries in the event of both main and standby power supply.

Remaining chains

At the time of writing, only the Japanese chain remained in service.

Accuracy and range

At close ranges to the baselines between the stations, a Decca fix has an accuracy of less than 100 m. However, as range from the stations increases, the lanes expand so accuracy reduces. The accepted accuracy is ± 1 nm by day for 95% of the time. The width of lanes also increases, and therefore the accuracy decreases, along the extended baseline outside the stations. However, other factors also reduce the accuracy of the system.

Sky waves

At the transmitted frequencies, sky waves return to earth after refraction from the ionosphere at a range of about 300 nm by day, and 200 nm at night. This means that outside these ranges, the system has to be regarded as inaccurate. At night, the system accuracy within that 200 nm range is also reduced to ± 5 nm (95% of the time).

Errors

Height error

The distance difference found from the phase difference is related to the slant ranges from the master and slave stations. If an aircraft is vertically above one of the stations, there will be a considerable error in the calculated position.

Night error

As already stated, sky waves interfere with the surface wave signal at long ranges. Around dusk and dawn the problem is at its worst.

Lane slip

Although the lane identification signals only occupy about half a second, the data flow in the receiver is disrupted for a whole second. A purple lane near the baseline is 385 yd wide. An aircraft flying across one of these lanes at 670 kts will cross that lane in less than one second, so lane slip at these speeds is quite probable. As we have seen, multipulse transmissions are able to reduce lane slip.

Static

Rain and atmospheric static can disrupt the signals.

Omega

Another hyperbolic navigation system operated on the phase comparisonphase comparison principle, but on VLF frequencies. This 'Omega' system was developed for the US Navy, because not only can the VLF ground wave signals travel vast distances, but they can also penetrate sea water to some depth. The Omega system was used by aircraft until the advent of satellite navigation systems made the complexities of the computers needed for Omega redundant.

Omega signals were sent from a series of eight ground stations around the world. Each station transmitted in turn using atomic clocks in a ten-second cycle, and their signals were maintained in the airborne equipment by locked oscillators. Baselines could be set up between any pair of these stations, and phase comparison of the signals from each end could produce position lines within lanes. Because the equipment was set up at start-up, and kept running thereafter, lane slip was avoided.

The signals from each station were of different pulse lengths. Each station transmitted the basic navigation signal on 10.2 kHz, 11.33 kHz and 13.6 kHz for lane identification and clock synchronisation. A computer in the airborne equipment used a complex database to correct the signals for the problems associated with the different propagation characteristics of the terrain over which the signals had come. At the transmission frequencies, the sky wave signals were attenuated less than the ground waves, and a combination was used to provide position lines. The computer database had to correct for ionospheric characteristics as well as the terrain.

The onboard computer could calculate and display position in latitude and longitude, or as range and bearing from a point. It could also provide distance, ETA and track to selected waypoints, and wind velocity, in the same fashion as an area navigation system.

Omega had advantages in its worldwide coverage, its long range of up to 10 000 nm, and its accuracy, down to within 1 nm by day and 2 nm by night. However, it had errors. These consisted of ionospheric anomalies, including one specifically at the polar cap caused by the magnetic flux, static interference from thunderstorms and geometric ambiguity when signals travelling the long way round the earth interfered with those with a shorter path. The errors involved, the complex calculations and detailed database required, the huge transmitter arrays and the high transmitter power required to achieve these advantages, were an incentive to switching off the system finally in 1997.

VLF range measurement

Not a hyperbolic system, but usually grouped with Omega because it used VLF signals from the Omega transmitters as well as from US Navy stations around the world, this provided position fixing in aircraft as well as the submarines and surface ships for which it was designed. It relied on the same accurate clocks as those shortly to be used for satellite navigation, as well as a similar standard of computer technology. It seems to have developed from an Omega lane identification system.

The technique involved measuring the actual phase of the carrier waves of the Omega or Navy transmissions, all of which started at known times on the accurate clocks. Practically, it relied on knowing the receiver's position, at least to within about 30 km, before it started moving. After that, locked oscillators guarded against temporary signal loss, and the phase measurement provided continuous position lines from the transmitters. Position lines from two transmitters, like those from UHF DME stations, could provide a fix, although for accuracy a third line was also needed to correct the clocks and oscillators.

The use of the system in aircraft was affected by clouds, such as cirrus types (see Volume 4), which contained ice crystals. These crystals in the path of the wave caused the signal to be lost. If temporary, the oscillators compensated, but frontal conditions, such as the ITCZ, were a greater problem.

The transmitters were particularly powerful, and the frequencies used were between 15 and 30 kHz. The system actually produced a more accurate fix than Omega, but needed a strong signal. This was easier to achieve with Omega because the predictability of the modulation allowed easier discrimination of the weak signal from the noise which could actually be stronger.

Decca summary

Principle	• Differential range by phase comparison • Master sends CW signal, slave delays, re-transmits harmonic, phase locked • Receiver converts signals to same frequency • Phase difference = hyperbolic line
Frequency band	LF (Harmonics of a basic VLF frequency 'f' around 14 kHz)
Harmonics	Master – $6f$, red slave – $8f$, green – $9f$, purple – $5f$

Indications	• Measures position within lanes ($\frac{1}{2}\lambda$ of comparison frequency wide) • Measured as decimals of lane width
Lane identification	• Once per minute, each slave transmits signals to give 1f. (Others switch off) • Compared with signals from master giving 1f • Gives decimals of zone – know which lane
Range	300 nm by day, 200 nm by night
Accuracy	• 95% accuracy of 1 nm by day, 5 nm at night, within range • Less accurate along extended baselines
Advantages	Accurate, not line-of-sight
Errors	• Height error – maximum above transmitters • Night error – sky waves possible beyond 200 nm • Lane slip – may reselect wrong lane after ident, max along baseline
V Chains	Old system as above. 'V$_1$' or 'V$_2$' depends on lane ident spacing
Locked oscillator	Reduces lane slip by continuing signals during ident
Multipulse	• Short pulse drives oscillators to be phase locked to transmissions • Extra 8.2f transmissions allow comparison at 0.2f for zone ident • Improves night range, gives automatic lane ident

Display	• **Early (V) systems** – decometers with lane ident meter
	  flight log – distorted map
	• **Modern systems** – computerised lat/long or along/across or grid
	  can feed EFIS

Answers to Sample Questions

Chapter 1 Examples

1.	100 MHz	**6.**	8797.6 MHz
2.	3000 m	**7.**	6 wavelenghts
3.	85.23 kHz	**8.**	109.53 MHz
4.	923.08 m	**9.**	135.01 kHz
5.	2.56 m	**10.**	208.48 m

Chapter 1 Sample questions

1.	c	**5.**	a
2.	c	**6.**	a
3.	a	**7.**	b
4.	b	**8.**	b

Chapter 2

1.	b	**8.**	d
2.	b	**9**	c
3.	c	**10.**	b
4.	d	**11.**	d
5.	d	**12.**	b
6.	c	**13.**	d
7.	d	**14.**	c

Chapter 3

1.	b	**6.**	a
2.	d	**7.**	c
3.	b	**8.**	b
4.	c	**9.**	d
5.	b	**10.**	a

Chapter 4

1.	c	**5.**	a
2.	d	**6.**	d
3.	b	**7.**	b
4.	c		

Chapter 5

1.	b	**7.**	b
2.	c	**8.**	c
3.	a	**9.**	c
4.	d	**10.**	c
5.	b	**11.**	d
6.	b	**12.**	c

Chapter 6

1.	d	**8.**	a
2.	b	**9.**	c
3.	b	**10.**	d
4.	d	**11.**	b
5.	d	**12.**	c
6.	a	**13.**	d
7.	c	**14.**	b

Chapter 7

1.	a. Parallel		b. 5° port
	c. 345°		d. 50 seconds
2.	a. 079° (M)		b. 135° (M)
	c. 315° (M)		d. 314° (R)
3.	d	**6.**	b
4.	b	**7.**	a
5.	a		

Chapter 8

1.	c	**8.**	b
2.	c	**9.**	b
3.	d	**10.**	c
4.	a	**11.**	a
5.	b	**12.**	a
6.	c	**13.**	b
7.	b		

Chapter 9

1.	c	**4.**	c
2.	c	**5.**	a
3.	d	**6.**	b

Chapter 10

1.	b	**5.**	d
2.	d	**6.**	b
3.	d	**7.**	b
4.	a	**8.**	d

Chapter 12

1.	d	**9.**	c
2.	b	**10.**	d
3.	c	**11.**	d
4.	c	**12.**	d
5.	d	**13.**	a
6.	b	**14.**	c
7.	d	**15.**	c
8.	b		

Chapter 13

1.	d	**8.**	a
2.	c	**9.**	b
3.	b	**10.**	d
4.	b	**11.**	c
5.	d	**12.**	a
6.	c	**13.**	d
7.	d		

Chapter 14

1.	c	**5.**	b
2.	b	**6.**	b
3.	c	**7.**	a
4.	d	**8.**	b

Chapter 15

1.	d	**7.**	d
2.	c	**8.**	a
3.	b	**9.**	b
4.	d	**10.**	c
5.	b	**11.**	b
6.	a		

Chapter 16

1.	c	**4.**	d
2.	b	**5.**	b
3.	a		

Chapter 17

1.	b	**6.**	c
2.	c	**7.**	d
3.	c	**8.**	a
4.	b	**9.**	d
5.	c		

Chapter 18

1.	d		**6.**	b
2.	c		**7.**	d
3.	a		**8.**	b
4.	b		**9.**	c
5.	d		**10.**	a

Chapter 19

1.	c		**8.**	d
2.	d		**9.**	d
3.	d		**10.**	b
4.	c		**11.**	b
5.	c		**12.**	a
6.	b		**13.**	a
7.	d		**14.**	c

Chapter 20

1.	c		**4.**	a
2.	b		**5.**	d
3.	b		**6.**	b

Chapter 21

1.	b		**7.**	d
2.	b		**8.**	d
3.	a		**9.**	c
4.	d		**10.**	b
5.	d		**11.**	d
6.	a			

Chapter 22

1.	d		**6.**	c
2.	a		**7.**	d
3.	a		**8.**	c
4.	d		**9.**	b
5.	b			

Index